The Darker the Night,
the Brighter the Stars

The Darker the Night, the Brighter the Stars

A Neuropsychologist's Odyssey Through Consciousness

Paul Broks

with drawings by Garry Kennard

CROWN
NEW YORK

Originally published in hardcover in Great Britain by Allen Lane, an
imprint of Penguin Press, a division of Penguin Random House, Ltd.,
London, in 2018.
crownpublishing.com

Library of Congress Cataloging-in-Publication Data
Name: Broks, Paul, author.
Title: The darker the night, the brighter the stars : a neuropsychologist's
    odyssey through consciousness / Paul Broks.
Description: New York : Crown, 2018.
Identifiers: LCCN 2018002690 (print) | LCCN 2018017108 (ebook) |
    ISBN 9780307985804 (ebook) | ISBN 9780307985798 (hardback) |
    ISBN 9780307985811 (trade paperback)
Subjects: LCSH: Cognition. | Brain. | Life sciences. | BISAC:
    BIOGRAPHY & AUTOBIOGRAPHY / Medical. | SCIENCE / Life
    Sciences / Neuroscience. | PSYCHOLOGY / Cognitive Psychology.
Classification: LCC BF311 (ebook) | LCC BF311 .B736 2018 (print) |
    DDC 153—dc23
LC record available at https://lccn.loc.gov/2018002690

ISBN 978-0-307-98579-8
Ebook ISBN 978-0-307-98580-4

Printed in the United States of America

Jacket design by Christopher Brand and Rachel Willey
Jacket photographs, from top: Fabrice Strippoli/Millennium Images, UK
(grainy face); © Bruce Peterson/Gallery Stock (anatomical textbook); Diana
and Endymion, Janneck, Frans (Diana and Endymion); Kevin Jones/EyeEm/
Getty Images (star trails)
Interior illustrations by Garry Kennard

10 9 8 7 6 5 4 3 2 1

First American Edition

For Sonja, a beautiful soul

Contents

All men think all men mortal, but themselves

—EDWARD YOUNG, *Night Thoughts*

*Now he has departed from this strange world a little ahead
of me. That signifies nothing. For those of us who believe in
physics, the distinction between past, present and future is
only a stubbornly persistent illusion.*

—ALBERT EINSTEIN, from a letter of condolence on the
death of his friend, Michele Besso

Prologue

T HIS WASN'T MY IDEA. GIVE THE READER A LAMP, THEY said, to lead them to the door. Pin a note to the door. Forewarn them it's a rambling, ramshackle house they're about to enter. I was more inclined to let you find your own way in and leave you to it. But, second thoughts.

This is not a conventional book and I think you should know what you're in for. What (I hope) you are about to read is a mix of memoir, neurological case stories, and reflections on life, death and the mind. I've thrown some Greek myths into the pot, and sundry other tales, some true, some not. Fact sits alongside fiction. Science tangles with myth. The fictional elements are, for the most part, easily identified. I don't have an autonomous subpersonality capable of seducing women in fluent French, for example, and I have yet to celebrate my one-hundred-and-fiftieth birthday. The case stories are mostly drawn from my experiences in clinical neuropsychology. Names have been changed and other layers of disguise added to preserve patient anonymity. There are two exceptions to the anonymity rule. One is the story of Pat Martino, the jazz guitar virtuoso, whose remarkable recovery from a near-fatal brain hemorrhage has long been celebrated in jazz circles. The other is Carla MacKinnon, filmmaker and sleep paralysis sufferer, whose short film about that strange condition, *Devil in the Room*, has been screened at festivals and medical conferences worldwide. As for the autobiographical stories, I have changed some names, and one or two other inconsequential details, but they are otherwise as true as I could make them. A few minor liberties have been

taken with the Greek myths, but, really, that's what you're supposed to do. All stories, fact and fiction, hang loosely around two
perennial questions: *What are we?* and *How should we live?* And,
throughout these pages, there is the echo of my wife's words in her
dying days: *You don't know how precious life is. You think you do, but
you don't.*

The book will make most sense if you read the first chapters
first and the last chapters last, but otherwise you should feel free
to skip and roam. Some chapters interlink, others don't, at least
not explicitly, but no doubt you will make connections I didn't
plan and haven't seen. The brain is a pattern-detecting device. It
finds shapes and designs and meanings all over the place. "Ah,"
someone said, "so it's a metaphor for the brain itself!" The way the
mind hangs together, she meant, logic linked to magic, dreams,
hopes, memories, broad vistas, blind alleys, and always a sense
of rolling along, hour by hour, of heading somewhere, even if that
somewhere, ultimately, is nowhere. Yes, I declared, you've hit the
nail on the head, though, in truth, it had never occurred to me.

There is no clear dividing line in the brain between inner imaginings and perceptions of the real, solid "world out there." Reality
and fantasy are built into the same neural circuits. I wanted my
stories to reflect that fact because, I believe, it rests at the core of
what it is to be human. That's why, as you find your way about this
rambling, ramshackle house of a book, you will encounter almost
as many gods, ghosts and mythic beasts as real-life people. It's also
why the neurological patients I have chosen to write about are so
often people who inhabit the twilight zones of the mind. You will
meet a man who believes he is dead but can tell you the story of
his life, and a woman whose life story has been erased. There's a
young man whose left hand has a menacing life of its own, and
an old man who doesn't know his left hand from his right or, indeed, his hands from his feet, or his feet from his ears, and women
plagued by foul-mouthed monsters of the dream-world as they lie
in their beds, wide awake but trapped in paralysis.

Observe the flow of patients through the neuro wards and clinics closely enough, and for long enough, and you will inevitably meet such people. Look into their disordered brains and you will learn something of the infrastructure of the self, yours as well as theirs. Look into their eyes, too, and see your own fragility.

So the lamp has led you to the door. Open it. Enter. Find your way along the gloomy hallway. Climb the stairs. At the top there is a door, slightly ajar. See? Push, and the door will open into a sunlit room, forever sunlit, regardless of the depth of the night.

PART ONE

—∞∞∞—

A Grief Observed

The Sunlit Room

B^{OOFFF} . . .

The oxygen machine exhales. It goes all through the day, all through the night. My wife exhales, like a sigh of resignation. It's six in the evening and she hasn't opened her eyes today, or spoken a word. This day, between her birthday and our wedding anniversary, is the day she dies. Yesterday the boys and I dabbed green tea on her lips and she smiled, but not today. Another long sigh. Her final breath? Not yet. There's another, and another. And then no more. The last is like the wash of a wave fading into sand. The oxygen machine is still breathing. I remove the wedding ring from my wife's dead finger, and box it in my fist. The machine exhales. I exhale. It scarcely missed a breath, this ring. I turn off the oxygen machine. Kate lies bathed in evening sunlight, the flesh of her arms already beginning to bruise with draining blood.

It was the autumnal equinox, September 23rd. The sun had crossed the celestial equator and our last summer was behind us. Perfect timing. She couldn't face another winter, she'd said. There was a full moon that night. I stood in the backyard. I took a slug of whiskey and I thought: What next? We had discussed *what next* a good deal that summer, knowing her death was imminent. "You'll be fine," she'd say, "I'm not worried about you." I had a lot going for me. It would be a release.

"And it won't be long now."

"Oh, that's all right then."

"But, I'll tell you something. You don't know how precious life is. You think you do, but you don't."

I couldn't argue with her. She was dying. What did I know? I look back on it now as a good summer, despite everything: painful, penetratingly sad, but without despair, and shot through with extraordinary moments of joy. It vindicated our decision. Precisely one hundred days before she died we were sitting in another sunlit room at the hospital. A doctor was telling us that the cancer had spread beyond all hope of containment. "How long?" Kate asked, and ventured her own estimate: "Six months?" But there was a pause before the doctor answered, "Perhaps." The best he could offer, the last resort, was another course of chemotherapy, which, if it worked, would extend her life by a couple of months at best. It would be the kind of chemotherapy that made your hair and fingernails fall out, and made you sick to your bones. We knew all about chemotherapy. And the chances of it working? "One in five." We didn't have to decide right there and then, the doctor said, the following week would do, but the disease was moving fast and treatment, if that was the choice, could not be delayed much longer.

We agreed, on the drive home, that it was not a decision to take on impulse. We would discuss it with Tom and Nat, our sons; we would weigh the pros and cons and do our best to make sense of the uncertainties. And in the days that followed we did those things. There was no agenda. Discussion came piecemeal over lunch on the patio, or watching the sunset up on the seafront, or in the quiet of the early hours, and we assembled the fragments forensically. It's your decision, the boys said. We'll support you in whatever you do.

My first thoughts, back there in the consulting room, had lined up pretty smartly against the idea of further treatment. Even as the doctor spoke, I was doing the existential equations. I factored in the probabilities alongside the pain and indignities, and I could see no good reason to intensify and prolong the suffering, which was already considerable. The end was inevitable and close now, treatment or not. Better to take what we could from the last days, not lose them to the ordeal of chemotherapy. If the treatment didn't work, which was likely, then it would just be adding insult to injury.

I kept those thoughts to myself at the time. If Kate was forming a different view, and I got the impression she was, then it was not for me to interfere. It was her life. And before long I began to see other sides to the argument. She had responded well to aggressive forms of chemotherapy in the past. Why not now? And why was the doctor being so conservative, so pessimistic, about the outcome? Oncology is not an exact science. They get these things badly wrong sometimes. I was given six months to live, you hear people saying, and here I am, five years on, fighting fit! So, I made the case for treatment. I said perhaps it was worth a shot. "I don't want to die with no hair," she said. Rational deliberation had little to do with it in the end. It came down to a *feeling of rightness*.

There are practical matters to deal with in the minutes and hours following a death. I called a doctor to conduct the certification, and a soft-spoken Ghanaian man showed up. I asked him if he could recommend an undertaker because, bizarrely in retrospect, I hadn't given the matter any thought. The doctor went on his way and I called the Co-operative Funeral Service and, while we were waiting for the undertaker, the boys and I took turns to say goodbye. I stroked her hair. When the body was removed, we—Tom, Nat and I, and Nat's wife, Rosie—ate some pasta and drank some wine. We talked about Kate. Her death felt, unexpectedly, like an accomplishment. It was a peaceful end, we agreed, a dignified one, and the suffering was over. I could not face spending the night in our, now just *my*, bedroom, so I laid a mattress on the floor in Tom's. I read Seneca's *Letters from a Stoic* for a while before settling to sleep, and I slept well. The next day, our anniversary, I took Kate's wedding ring to a jeweler for resizing. I'd promised her I would wear it for the rest of my life.

In the days that followed there was the funeral to arrange, and details to gather for the Registrar of Births and Deaths, who, when

I got to see him, told me he was sorry for my loss, a phrase that must pass his lips fifty times a week, and then he gave me an old-fashioned fountain pen to sign some forms. Then there's the funeral, and that's it. A life concluded; a death documented.

Then the memories started pushing through. Doors opened into unexpected rooms. Through this window, a crisp winter morning, through that, a summer's afternoon. Fragments of childhood swirled up like leaves in a flurry. Schooldays. Work. The early years with Kate. I opened the back door and there we were, standing in a downpour. The scent of hard rain on dry earth. Soaked to the skin. Alive. The images were involuntary and spasmodic, as if my brain were trying to gather threads of meaning without much involving "me," churning the memories, poking and probing. Reconstructing. *Who are you? What next?*

What next? No idea. I was wandering through a mist, not knowing what to expect when the sun burns through. *When I'm gone, just get on and do whatever you must.* But what? Sell the house, she said. Pack in the job. Move to another town. Find another woman. Anything. *I'll be just a memory.*

I decided to follow Kate's advice and retire from work at the earliest opportunity. *You're getting stale.* I was. *You've no appetite.* True. *Let go.* She had it all figured out. I could use her life insurance money to pay off the mortgage, and, within a couple of years, I'd be eligible to apply for an early retirement package, which would give me a small pension to live on. So I found myself entering a branch of the Cheltenham & Gloucester Building Society, briskly signing a check for ninety-six thousand, four hundred and eighty-eight pounds, forty-three pence, and going back out into the street with a tear running down my cheek. You get those stabs of absence to the gut when you least expect them. Eighteen months later I resigned my university post and got on with the things I'd much rather be getting on with. Walking the moors. Country pubs. Football. Reading. Loafing.

Believe me, I'm a good loafer, but my brain wouldn't rest.

The Wooden Sword

M<small>Y FIRST ENCOUNTER WITH THE BRUTE FACT OF DEATH</small> was this. I see myself in the backyard, a small boy, watching his father, sleeves rolled, chipping away at a length of wood with a pocketknife. He was forever hammering and fixing things and making stuff. Sometimes he turned his hand to toys: bows and arrows, catapults and carts. On country walks he cut whistles from birch branches. He once turned an upright piano into a dropleaf table. Now he's putting the finishing touches to a sword. I remember the whiteness and the sappy scent of the stripped wood. I loved the sound it made when I swung it against the gatepost. *Thock!* Job done, Dad goes back inside the house.

Three boys appear, brothers new to the neighborhood. They want the sword. "Ah, go on," they say. "Lend it to us. We'll bring it back tomorrow." I don't want them to have it but they take it anyway, and I watch, mute, as the three of them scrap over first rights to the trophy, and then they're off, laughing and cursing, and I feel bad because I've let my dad down. I tell him they've promised to return the sword tomorrow, but he shrugs and says no, that'll be the last I'll see of it.

That night the boys' house burns down. They, their mother and two sisters all perish. The father survives. Next morning I go to look at the smoldering shell of the building, at the glassless windows and blackened bricks. There are television news cameras. A dead, damp smell hangs over the place. In the following weeks, I develop a fear of fire so bad I can't tolerate the fire my dad builds in the kitchen grate each morning. "Look," he says, striking a match, "nothing to be afraid of." But it's as if he's struck the match inside my stomach. I'm taken to see the doctor, who also strikes a match.

I had already considered the dread possibility of my parents dying, but now I start to contemplate my own death. I know about Heaven and Hell and expect to go to Heaven, so why worry about dying? Heaven is a nice place where good things happen forever. It's where God lives and it would be a happy place to be, a fairground of a place, and the happiness would never end. But now the wraith of a different fate is lurking in the shadows of my brain: nonexistence. Maybe the three brothers had gone to Heaven but, then again, maybe they hadn't gone anywhere. Perhaps their bodies had been burnt to cinders and that was that.

It said in the local newspaper that a paraffin heater toppling over and setting light to a mattress probably caused the fire. I see the brothers in their pajamas. They're playing pirates or something, charging around and jumping on their beds and rolling around, and their dad is shouting up the stairs at them to keep the noise down. The paraffin heater is hissing softly. They ignore their dad. Then one of them grabs the wooden sword and swings it wildly and it catches the top of the heater and sends it tumbling against a bed. It set me wondering. If I'd refused them the sword, if I'd been stronger, perhaps the fire would not have happened and the brothers and their mother and sisters would still be alive. But nobody knows for sure what really happened, only God, because he watches everything. He saw what started the fire and he watched them all burn.

Know Thyself

A POLLO, THE GREEK GOD OF HEALING AND DISEASE, MUSIC, logic and light, founded a temple on the slopes of Mount Parnassus. He had journeyed from Crete in the form of a dolphin, and Delphi became the name of the sacred site upon which the temple was built. Inscribed in the forecourt was an injunction: *Know thyself*. For centuries thereafter Apollo channeled prophecies through the Oracles at Delphi, a succession of priestesses, called Pythia. It was a Delphic oracle that inspired the great thinker Socrates to pursue his "divine mission" in philosophy. His friend, Chaerephon, had asked the oracle if there was anybody alive wiser than Socrates, and "None" was the answer. Socrates was intrigued because, he said, if there was one thing he knew, it was that he knew nothing. Apollo could not be wrong, though, so perhaps it was his own understanding of wisdom that required examination. What was it that he, Socrates, had that set him apart from the politicians, the poets and the craftsmen who seemed to be blessed with more knowledge and talent? He made it his mission to solve this riddle and, in so doing, set the course for the future of Western philosophy and science. It turned out that what he had in abundance was the capacity to question, to doubt and to reason. He also had the wisdom to see that the pursuit of truth was more important than the pursuit of material wealth. Truth and the perfection of the soul through the cultivation of virtue were the foundations of a good life. This required continual self-scrutiny. The unexamined life, he said, was not worth living.

The "good life" was also a central concern of the Stoic school of philosophy, founded in Athens by Zeno of Citium around 300 BCE, a hundred years or so after the death of Socrates, and flowering three hundred years later in the work of the great Stoic philosophers of the Roman era: Seneca, Epictetus and Marcus Aurelius. For the Stoics, a life worth living was one lived in accordance with nature, which meant not so much living in harmony with the natural world but, more particularly, living in line with our nature as human beings. So the question *What are we?* precedes *How should we live?* Now, for the Stoics, we are, above all else, rational beings, and should strive to lead our lives accordingly.

Friedrich Nietzsche, the nineteenth-century German thinker, who referred to himself variously as the first Immoralist, the Anti-Christ and the Annihilator, didn't have much time for the Stoics. He thought their injunction to live in accordance with nature was vacuous, but his own Theory of Eternal Recurrence contains something of the spirit of Marcus Aurelius:

> *What if a demon crept after you one day or night in your loneliest solitude and said to you: "This life, as you live it now and have lived it, you will have to live again, times without number; and there will be nothing new in it, but every pain and every joy and every thought and sigh and all the unspeakably small and great in your life must return to you, and everything in the same series and sequence—and in the same way this spider and this moonlight among the trees, and in the same way this moment and I myself. The eternal hour-glass of existence will be turned again and again—and you with it, you dust of dust!"—Would you not throw yourself down and gnash your teeth and curse the demon who thus spoke? Or have you experienced a tremendous moment in which you would have answered him: "You are a god and never did I hear anything more divine!"*

So instead of imagining each day to be your last, as Marcus Aurelius counseled, consider the opposite: the possibility that each and every day is destined to be repeated, in precise detail; that your whole life will roll out over and over again for all eternity. Either way, you would strive to make it a good one.

The Stoic recommendation to live life in accordance with our rational nature is all well and good. The faculty of reason is a distinguishing feature of the human mind, but we have other distinguishing features, and if we are to heed the maxim inscribed in stone at the Temple of Apollo—*Know thyself*—then we must also take into account intuitions, dreams and imaginings. More than that, in the Socratic spirit of questioning everything, we should go further and doubt whether it is actually ever possible to know one's self. We should doubt even that there are such things as selves.

I once gave a public lecture in a provincial town hall. At the end of my talk a woman rose from her seat and strode toward the front. "You smug *know-it-all!*" she shouted, "I want to shake you by the lapels!" One of the organizers stepped up ready to intervene, but the woman was already making for the exit. "*You* might be a soulless machine," she said in a parting shot, "but I most certainly am not." She was upset because I'd rounded off the talk with a nihilistic flourish. I'd said that studying brain function and working with brain-damaged people had led me to the view that the inner sanctum of the self is a void. There is no inner sanctum. Science has done away with the soul. There's no ghost in the machine, just a machine. But the soul's secular cousin, Self, doesn't really stand up to scrutiny either, if by self you mean some immutable inner essence, a fixed, observing "I" that follows us down the years, that is, in fact, "us." Oh, and by the way, free will is also an illusion. Altogether, our deepest intuitions about what it means to be a person are false. The human brain is a storytelling machine and the self is a yarn it spins. That's it. Nothing more. The story is all. Blah, blah, blah. I've reeled out this litany of self-annihilation ad nauseam over the years. Sometimes I feel like shaking myself by the lapels.

Boy Sleeps, Man Wakes

THERE'S A BOY, NINE YEARS OLD, FLYING THROUGH THE cosmos. He doesn't comprehend the vastness of interplanetary space. Or the immensely greater vastness of interstellar space. Or the inconceivable volumes of intergalactic space. So the planets and stars and galaxies shoot by at the speed of thought. Vooom! It's a journey he often makes before falling asleep. His destination is the edge of the universe, which is a featureless gray wall, infinitely deep, infinitely tall. On arrival he contemplates the wall then turns to look back at the stars and galaxies gleaming in the distance, then turns again to look at the wall. He stares hard and runs his hands over the smoothness beyond smoothness of its non-texture. The idea of nothingness beyond the wall is ungraspable. He has reached the limits of his imagination as well as the boundary of the cosmos, and there are butterflies in his belly.

Sleep won't come. Thoughts are running like rats through his head and a shadow on the far wall of the bedroom unsettles him. He wonders: Why is there anything? Why is there *stuff*? He wonders: Where was I before I was born? He stares at his hands, deep into the ridges of the skin. *I'm still here.*

The shadow moves again, not the blurred and elongated wings of the Lancaster bomber, or the half-painted Joan of Arc, or the Spitfire or the Black Prince. His eyes follow the Airfix models across the mantelpiece, to Napoleon Bonaparte. He alone was capable of movement. Magical movement. Not much, but some. A cocking of the head. A shuffling of the feet. A stooping of the shoulders. Or so it seemed. He was nothing to start with, Napoleon, just a set of gray plastic fragments in a box, but in the process of assembly—arms, legs, torso, face—a malignant presence grew. When the painting was done—blue greatcoat, white breeches, black boots—the presence filled the room.

The boy gets out of bed, walks across the linoleum and pisses into a bucket. He goes back to bed, never taking his eyes off Napoleon. Great pellets of rain start rattling the tin roof of the shed across the yard.

Next morning is bright and sunny. It's a Saturday. There's pop music playing on a radio somewhere. The boy has been lying on his bed, staring for a long time at the Airfix models on the mantelpiece: the Lancaster bomber; Joan of Arc; the Spitfire; the Black Prince; Napoleon Bonaparte. Bonaparte stares back. He gets up from his bed holding the Emperor's gaze. He goes over and grabs him and walks steadfastly down the Emperor's tight curve of stairs to the kitchen, and he throws him on the fire and watches him melt. The fire doesn't frighten him anymore.

Man wakes.

The night was teeming with dreams. He stares at the ceiling. The sleep-stiff cogs and gears of thought crank up. To begin with we are nothing. Sperm meets ovum. The fertilized egg divides, and the divisions divide, and go on dividing, and a fetus develops. Arms, legs, torso, face slowly assemble. A baby is born, grows into a child, and somewhere along the line a flame ignites. Flesh and bone make magic. Then, sooner or later, the flame goes out.

He fumbles for the radio on the bedside table, then stares at his hands, deep into the ridges of the skin. *I'm still here.*

Monday, October 17th

Slept well enough but felt a bit ragged due to a couple of drinks over the usual last night.

Lay in listening to the radio. Start The Week, *with Richard Dawkins, Rabbi Jonathan Sacks, and the physicist Lisa Randall, all plugging their latest books. Sacks came over as smug and incorrigible. Dawkins sounded weary of him. The Chief Rabbi spoke blithely of Christianity as a "right-brain religion" translated into a "left-brain language." He made a portentous closing statement: "Without God, there is no hope." In between he was claiming that only religion could answer the questions: "Who am I? Why am I here? How should I live?"*

10 min late for a 10 o'clock tutorial, students waiting blank-faced outside my office. Lectures all afternoon. Walked home via Devil's Point. Sat half an hour on the bench where I used to sit with Kate, watching darkness descend over the turbulent water.

Going down with a cold. Thought I deserved a medicinal whiskey. Just the one. A large one though.

My grandson was due to arrive in the world today, but no new baby by close of play.

AN EMPTY THEATER. On stage: two old men sitting in armchairs. An AK-47 assault rifle is propped against one of the chairs, and there's an electric guitar against the other. The objects are the old men's offspring, because, it seems, the man on the left is Mikhail Kalashnikov, who designed the rifle, and the man on the right is Les Paul, who designed the guitar. A vodka bottle stands on a low table between the two of them, the brand named in Kalashnikov's honor. Outside, through the window, it could be Izhevsk or it could be New Jersey. Kalashnikov stands, picks up the guitar and pulls it to his bemedalled breast. It's heavier than he thought and his arms sag. But, with a clowning face, he nods at Les, then at the AK-47 and then again at Les. No words are exchanged. Les gets up and walks to the front of the stage. He shields his eyes against the lighting and says, "You'll have to remind me. I can't remember her name. What was her name? I'm getting a little forgetful." I'm watching from the stalls. A woman in the seat behind leans forward. She whispers, "*You* might be a soulless machine, but I most certainly am not."

The meaning of the dream seemed clear at the time.

Wednesday, October 19th

Nat phoned at 2:20 a.m. relieved and happy that the baby has arrived. Quick delivery at 1:03. Rosie is fine. At lunchtime Nat sent a picture, baby settled in his cot after a bath. Beautiful. This is the anniversary of the day I met Kate.

Baby Harry was one of around 370,000 new human beings launched into life on planet Earth on 19 October 2011. The same day that he and his cohort were taking their first breaths and blinking at the blooming, buzzing confusion of it all, roughly 155,000 people took their last and slipped into oblivion. Entrances and exits. We are born; stuff happens; we die. Why? No reason. The rabbi asks, "Why am I here?" but it's a question loaded with the assumption that there is in fact some preordained, cosmic purpose to life. No, we are here by way of the blind physical forces of the universe, the evolution of life on Earth, and the contingencies of human history. We're here because we're here because we're here, as the soldiers sang in the First World War.

We have no say in being born, and the fact that we exist at all is a matter of pure and improbable accident: that particular mother; that particular father; that particular egg from the million a woman is born with; that particular sperm from the 250 million released in an average ejaculation. A vast cascade of events, big and small, determines our development from conception to the present moment.

My father, a Latvian refugee, arrived in England at the port of Hull in 1947. He and his compatriots were taken to Shropshire to work on the land. It was there he met my mother one day when she was out cycling in the countryside. She had a puncture; he fixed it. So, if I owe my existence to the Second World War, and to my father's status as a displaced person, I owe it, no more nor less, to a nail left lying on a country road.

My wife and I met because I missed my bus and turned up too late for a date with someone else. You look a bit lost, she said. Our children and grandchildren owe their existence to my unpunctuality, as well as to the nail on the road and the Second World War. Et cetera, et cetera. We're here because we're here because we're here.

"SAME AGAIN?" THIS is my friend Rob.

"No, I've got a date."

"Half?"

"Go on, then."

A glass of beer. A missed bus. Another throw of the dice.

THE ODDS AGAINST me being "me" and you being "you" are incalculable. But here we are. We don't choose our parents or the historical period and culture into which we are born. Sex is genetically determined as is, to some degree, sexual orientation. Nor are we blank slates when it comes to temperament or potential for achievement. I could never have been an Olympic athlete, a chess grand master, or an operatic tenor. My limbs, my lungs, my brain and my vocal apparatus are not built that way.

In Burma my grandfather came close to being eaten by a tiger, or so the story goes. I like to think it's true and sometimes imagine a parallel history in which the tiger sprang and swallowed all possibility of my future existence. It's somehow less troubling to think you might never have existed than trying to grasp the idea of ceasing to exist at some unspecified instant to come. But the tiger didn't spring. Corporal Fred Harris returned from the Burmese jungle to work in a coal mine, and to sire three sons and a daughter, too, who, by virtue of Adolf Hitler and a nail on the road, married a Latvian émigré. And here I am, like all of us, a bundle of meat and bone thrown naked into the world. But what do you do once flung? You have to do something. How best to live? Life may be short, but it's long enough to make terrible mistakes.

I didn't plan to but time and again during those last months with Kate I found myself reading the Stoics, mostly Seneca and Marcus Aurelius. They spoke plainly, and with a remarkably modern voice. Seneca was born in the year 4 BCE, which, according to historical analysis, is around the same time as Jesus of Nazareth. (And what a contrast there is between the measured, erudite Roman and the Nazarene Son of God with his magic and his hectoring and his wild ideas.) Seneca's essay *On the Shortness of Life*, which I'd first read in my twenties, made a sharper impression in my fifties, an age closer to the author's and with life's dusk a-gathering. At times, it felt almost as if I were conversing with a friend, and, unlike his holy contemporary, he was not promising eternal life. Quite the opposite.

> *You are living as if destined to live for ever; your own frailty never occurs to you; you don't notice how much time has already passed, but squander it as though you had a full and overflowing supply— though all the while that very day which you are devoting to somebody or something may be your last. You act like mortals in all that you fear, and like immortals in all that you desire.*

Time, for the Stoics, is life's most precious commodity. Be mindful of it. Don't waste it. Keep death in view. Savor the past (if you can), anticipate the future, but above all grasp the present. Live immediately. Avoid, or tame, the negative emotions: fear and anger, jealousy, hate and sorrow. Aspire to tranquillity. Make the present more habitable. You have a brain. Embrace reason. You have a heart. Show love. Feel gratitude. Respect your fellows. Honor your social duties. These things are at the core of what it is to be human. *Keep death in view, always.* Keep death in view and, when it comes, at whatever age, be content to have lived a good life. Die happy, or, if not happy, at least without serious and irredeemable regret for the life you have lived.

Sunday, October 23rd

Early morning dream of Kate drowning in a pond part-covered with ice. So I've seen her consumed by fire (that hallucinogenic dream a couple of days after she died) and now ice. Each time she was knowing and unconcerned.

Drove over to Somerset. Rosie was cooking an old-fashioned Sunday roast.

Wonderful to hold Harry for the first time. He seems a strange mix of settled, content but also troubled-looking, as if genuinely perplexed to find himself in the world—a "what's all this then?" sort of look. The girls were a whirlwind of energy as usual. Champagne and a Star Trek toast to Harry—"Live long and prosper!" Tom returned from his wedding photography work around ten and we three boys drank whiskey and watched the football.

Bedtime reading—Stephen Hawking. Something to think about: the idea of the "indeterminate past" of the quantum world. Is the "personal past" also indeterminate? In what sense does "the past" exist at all beyond our constructions of it? Could it be

*indeterminate in a deeper sense just like the past of the quantum
world is really, truly not fixed?*

THE BEGINNING OF time and the end of time are, like the edges of
the universe, unfathomable. Equally unfathomable is the thought
that there is no beginning or end. The boy (I will claim him as
me, at least for now) sometimes pictures himself sitting in a clear
plastic bubble. It's a time machine. There's a lever. You push it to
go forward and you pull it to go back and there's a digit counter
showing the date. So there he is, sitting in the bubble, and he yanks
the lever back, hard, and the world outside the bubble is a blur of
Spitfires, penny-farthings, knights in armor and wheeling ptero-
dactyls. Then it's a barren, volcanic landscape, and then there's
nothing but space and stars and silence, because this is a time before the Earth was formed. Then what? Was there a time before
that, before the stars, when there was nothing at all? He watches
as, one by one, the stars are sucked back into the void until, finally,
he's left sitting in the bubble with the years rolling endlessly back
on the counter, black space all around, nothing happening.

The journey into the future goes like this. You whiz through a
world of robots and flying cars and gleaming skyscrapers thrust-
ing up through the clouds, and silver-finned rockets flying to the
Moon, but, soon enough, all this disappears. The Sun gets old and,
in its death throes, expands and swallows the Earth, and all the
other stars get old and die. You sit in the bubble in the dark, dead
universe watching the years rolling on forever and ever. Nothing
happening. Everything comes to an end in the end, and the end is
endless.

Big Bang, Little Whispers

WHEN WAS THE LAST TIME I SLEPT THROUGH THE NIGHT? I woke at four in the morning and tuned in to the World Service, as usual, to dull my brain, but this time it kept me awake. There was some physicist talking about the origins of the universe and how physics can explain how something, that is, the universe, can pop into existence out of nothing. But surely there has to be the *potential* for something to happen, and potential would be something rather than nothing, wouldn't it? And if so, where does that come from? Or is that a stupid question?

Clearing out the utility room I found some old photos, boxed and carefully cataloged: Oxford, '80–'83, Leeds '87–'91, and so on. Kate must have sorted them in the last months. I seem to remember she said she would, but didn't take much notice at the time. There's a picture of her in full bloom, naked and pregnant. Others from long-gone Greek summers and Lake District winters. How fresh and pretty she looks.

In the afternoon I went out for a walk on Dartmoor, a soggy, blustery trek to the prehistoric settlement at Merrivale and the tors beyond, and I had this thought sitting atop King Tor: 13.8 billion years ago there's a Big Bang and *all this comes from nothing*! Then thoughts about the thought: a thought is something reducible to electrochemical pulses in the circuits of my brain, so a representation of the Big Bang and all that *something from nothing* is coded in my neurons, in ways mysterious to *me*. And then this thought about that thought: there's another kind of something from nothing going on here because there's also a spark of consciousness in the mix, which is, in effect, all that something from nothing out there—the moors, the sky, the stone circle in the valley, the whole something from nothing cosmic shebang—*thinking about itself*, conscious of itself, through the machinery of me, a bundle of neurons and sense receptors perched, for no particular reason, on a slab of rock. Big Bang, and, billions of years later, little whispers. Little whispers of awareness. The universe talking to itself. Perplexed. Bemused.

I listened to the football commentary on the drive home. Wolves v. Albion. Wolves lose 2–0. That's five defeats in a row. I can't bear to listen after the second goal goes in and had to avoid *Match of the Day*. Why does it bother me, even slightly? I was getting engrossed in games even through the toughest times with Kate, when, if ever football didn't matter, it was then. But still, you do.

That Thing You Do the Time With

The great topmost sheet of the mass, that where hardly a light had twinkled or moved, becomes now a sparkling field of rhythmic flashing points with trains of traveling sparks hurrying hither and thither. The brain is waking and with it the mind is returning. It is as if the Milky Way entered upon some cosmic dance. Swiftly the head mass becomes an enchanted loom where millions of flashing shuttles weave a dissolving pattern, always a meaningful pattern though never an abiding one; a shifting harmony of subpatterns.

—CHARLES S. SHERRINGTON, *Man on His Nature*

"FRANK, WHERE'S YOUR NOSE? WHERE'S YOUR ANKLE? SHOW me your ear."

The younger man is saying these things to the older man sitting opposite. They are facing one another, just close enough to touch. Frank is concentrating hard, giving each question careful thought. He responds by pointing in turn to his right shin and his left forearm and then by patting the top of his head. "Where's your shoulder?" He points to his left knee.

I'm sitting with a group of students watching a video of Frank and my younger self performing a familiar routine. The old man is long dead and it occurs to me that, with the passage of time, twenty-two years, every molecule of my younger self has been replaced. The body's tissues are constantly regenerating. Old cells are discarded and new ones grow to take their place. Neither of those bodies has survived. But I'm still here. I shift my eyes from the screen, look close at the palm of my hand, deep into the ridges of the skin. Yes, I'm still here.

Now we're watching Frank and me sitting side by side at a white-topped table. I reach into a bag and place a clock on the shiny surface. He watches without comment as various other objects are set in a row before him: next to the clock is a fresh strawberry, next to that a toothbrush and, finally, a banana. The banana gets a laugh. "What's this?" I say, pointing to the clock. "Oh it's a . . . you know . . . oh what is it called . . . that thing, you know, that thing you do the time with. I can't remember. I know what it is." Each of the objects elicits the same kind of response. "I know what it is," he says of the strawberry and the banana, "you eat it." He picks up the toothbrush and mimes a brushing motion in front of his bared teeth.

My former, long-dismantled self is sitting with long-departed, long-disintegrated Frank in a windowless basement room in St. James's Hospital, Leeds. The year is 1989 and this is my first job as a neuropsychologist. Prior to this I've spent six years doing clinical psychology training and doctoral research, following which a couple of years as a scientist with a pharmaceutical company, researching the effects of drugs on memory. So by the time I get to St. James's I'm familiar with the landscape of the brain and have a working knowledge of its disorders and frailties: the strokes, the degenerative diseases, the epilepsies, the traumatic injuries, and so on, but that hasn't prepared me for the daily grind of life in a general hospital. The variety of referrals is bewildering. All those years of study and research and, I realize, I still don't know very much. It still feels like that. Whatever I am, and whatever I've done, it's always felt like I've blagged it and someone, sooner or later, is going to find me out.

I was something of a misfit at school. My biology teacher, Mr. Bingley, wrote in an end-of-term report: *Resists learning as a deliberate policy.* He told me I was a nihilist and that he expected me to be hanging from a rope sooner or later. Mr. Bingley and

I got along quite well, which was not the case with Mr. Clamp, the physics teacher who threw me out of class one day for asking, in innocence, if thoughts consumed calories. It was a violent throwing-out, with him left clutching tufts of my hair. As to whether thoughts consume calories, I still don't know the answer. It's not straightforward. The working brain consumes calories, and thoughts require brain activity, so the act of thinking consumes calories. But thoughts? What is a thought?

Kidderminster Harriers, the local football club, signed me up, and I was doing well in the youth team. Then Wolverhampton Wanderers offered me a trial. They were a top team at the time, one of the best in England, and I was a fan. So, I didn't pay much heed to the Bingleys and the Clamps. I was going to be a professional footballer. The letter of invitation, signed by Joe Gardiner, former Wolves player and trainer, then scout, remains by far the most exciting letter I have ever received from anyone. Joe Gardiner was a man of few smiles and, at the end of my trial, he took me aside and told me, unsmilingly, that I wasn't going to make the grade.

I idled through my last year at school, left with meager qualifications, and went to work in a carpet factory with my uncle Ray, rolling out, measuring and cutting lengths of Axminster and Wilton on eight-hour shifts, six-to-two, two-to-ten. It was physically undemanding and gave me time to think and, when work was slow and the bosses weren't around, to read. Mostly, I read stuff about the mind and the brain: *The Pelican History of Psychology*; George Miller's *Psychology: The Science of Mental Life*; Aldous Huxley; R. D. Laing; J. Z. Young; Oliver Zangwill. As the looms clattered on the floor below, my brain was doing its own weaving.

TWELVE MINUTES AND twenty-seven seconds into the film we catch a glimpse of Frank's wife sitting in the background. I don't remember her name. She is still and expressionless like a waxwork.

The clock keeps real time. It shows 3:41 and, five minutes later, 3:46. Frank's wife doesn't move at all. The clock keeps real time in this edit of the film, but there's another version where it doesn't. Five minutes shrink to two. Time goes backward and forward. In a third version the clock is absent altogether and so is Frank's wife.

"He's so sweet," one of the students says. "He reminds me of my granddad." He reminds me of mine, too, and that generation of granddads who wore a suit and tie to a hospital appointment.

I ask him to remember to do something. I've set a timer to sound an alarm, at which he is to remind me to telephone my wife. Ten minutes go by and the bell rings. He ignores it. I look at him expectantly. He looks at me confused. I remind him there's something he's supposed to remind me about. "Ah," he says. "Your lady." The word *wife* eludes him. "You've got to telephone your lady."

I pause the video. We've seen enough, or I have. I turn to the students.

"So what's going on here?"

Silence.

"Any ideas?"

Silence.

"Wild guess?"

"Alzheimer's?" someone says.

"Is it a vascular dementia?" someone else ventures.

"It's not a dementia."

"Stroke?"

"You're getting there."

I tell them to forget about the brain for now. Focus on the mind. What are Frank's *functional deficits*? What can't he do? Psychology is never enough to confirm a neurological condition, I explain, but in Frank's case, the diagnosis was unexpected and psychological assessment was instrumental in identifying the problem. I tell them we'll discuss the case next week but they should mull it over in the meantime and see what they can come up with.

I remember seeing Frank off at the end of the session. His wife gave him a dig: "Frank Slater, you don't know your arse from your elbow." "No, love," he said.

To the students: "Clue: what's the most striking thing about Frank's presentation?"

Silence.

"He doesn't know his arse from his elbow."

Silence.

At the end of the lecture I resolve never to show the video again. I'm tired of seeing my video-self getting younger, while the flesh-and-blood version grows relentlessly older. I make this resolution every year.

In memory, I follow Mr. and Mrs. Slater through the door of the windowless basement room, along busy hospital corridors and out into the sunshine of a Yorkshire summer. They go their way, I go mine. I have taken off my tie. I am driving home. What was the car? Can I reconstruct the route? Not sure I can, precisely, but more or less. I drive through mazy streets of redbrick back-to-backs and out onto the York Road, on through suburbs whose names are lost to me, to Garforth, an old mining town six or seven miles away, to my own redbrick terrace house. Inside I find my "lady." In this time-trip she wears blue jeans and a flame-red top. Toddler Tom jog-toddles to greet me. Blond baby hair, blue dungarees. Nat's in the backyard kicking a football.

Stardust

I T TAKES ROUGHLY SEVEN BILLION BILLION BILLION ATOMS TO build a 70kg human being. Adjust this figure according to your weight. The bulk of you (93 percent) is made up of oxygen, carbon and hydrogen, with nitrogen, calcium and phosphorus atoms accounting for most of the remaining 7 percent of your mass. Hydrogen has been around since the Big Bang, but the other elements are spewed out from the fusion factories of dying stars.

Atoms get recycled. Your atoms were once the atoms of other objects and people. You contain atoms that once were the atoms of birds and trees, oceans and clouds. It is statistically almost certain that, in the course of your life, you will ingest atoms once breathed by Hitler and Buddha, Newton and Socrates, atoms that once formed the body and the blood of Christ.

This is how we are assembled. Atoms combine to form molecules, which in turn combine to form cells, the fundamental units of life. Cells group together to form tissues. Tissues assemble to form organs: heart, lungs, spleen, stomach, pancreas, kidneys, liver, skin, bones, intestines, bladder, sex organs, eyes and brain. You are a complex machine. Your heart is a pump. Your stomach is a fuel processor. Your brain, an electrochemical supercomputer (of sorts), is the machine's control center, regulating internal processes and guiding outwardly observable movements of limbs, torso, head and face. We also think of the brain (it thinks of *itself*) as the organ of mind. Everything you do, say, think and feel can be traced back to brain activity. It's debatable whether brain function alone is sufficient to account for mental life (there are brain/body and body/world interactions to consider) but it is most certainly *necessary.* No brain, no mind. When he was about five years old, my son Tom asked me, *Daddy, what's inside your head?* A brain, I told him. *What's inside that?* Just brain stuff, I said. *What's that like?* It's a bit like porridge. *Oh,* he said, and carried on pummeling his Play-Doh.

Neuropsychology is the study of the relation between brain and mind. We know what the brain is. It's an organ located in the head. But what is the mind? As a neuropsychologist, my approach was a practical one. For clinical purposes, I treated the mind as a confederation of processes: perception, emotion, reason, language and memory, each faculty with its own further subdivisions. The mind is not a monolith. It is, to use the jargon, *modular.* You sometimes find severe malfunction in one mental domain alongside normal operation in others. This is because, despite considerable overlap and interconnection, different brain systems serve different psychological functions. For example, reading impairments (alexia) are neurologically dissociable from writing impairments (agraphia). Some people lose the ability to read but can still write, and vice versa. Likewise, memory for words is distinct from memory for faces, and memory for places relies on yet other circuits. And so on.

So there we have the brain and the mind, but what of the self and the soul? This takes us a step further into the storytelling zone, into the realms of imagination. The Victorian artist Samuel Palmer said a picture was "something between a thing and a thought." The same can be said of a person. Paintings and human bodies are physical objects that can be weighed, measured and analyzed in different ways, structurally, chemically and so on. But in each case the material form is only a part of what we see. When we look at a picture, Palmer's *Cornfield by Moonlight*, say, it's not the paper, the paint, the ink and varnish we see. It's not just the depiction of a man and his dog in a wheat field under the light of the waxing crescent moon and the evening star. We are transported beyond the physical and the literal into the numinous, into a world of gods and spirits. Something similar happens when we look at one another. We can't help it, even if we don't believe in gods and spirits.

Sometimes I saw Kate even though she wasn't there. Nothing spooky. I would catch something of her smile in another woman's smile, or hear her in someone else's voice or laughter. It didn't happen often but when it did it was oddly comforting. Once, I followed a woman to the end of the street because she had Kate's figure and walk and wore the kind of clothes that Kate would wear. For a minute or two I had the sense that she was still alive. I could catch up with her and we would carry on as normal. The woman turned left at the end of the street and I turned right, and I went home, spirits lifted, just a little.

Autotopagnosia

Dr. J. M. Carter,
Consultant Neurologist,
SJUH

27 July 1989

Dear James

Re: Mr. Frank Slater, dob 09/04/14; 22 Bartholomew Road, Leeds

Thank you for referring this very pleasant 75-year-old gentleman for neuropsychological assessment. Full report attached.

Key points: Mr. Slater presented with a 9–12 month history of cognitive decline, characterized by forgetfulness, word-finding difficulties and intermittent episodes of confusion. I found evidence of some moderate decline of verbal intellectual capacity, in the context of intact nonverbal function, although this is hard to evaluate given his dysphasia. The same verbal/nonverbal disparity was also found in the memory domain.

The overall pattern does not fit with a dementia of the Alzheimer type. My hunch is that the underlying cause is vascular. Most strikingly, Mr. Slater is showing clear signs of autotopagnosia [body-part misidentification] which probably implicates the posterior left hemisphere. This is something you might want to investigate further.

Best wishes,
Yours sincerely,

Paul
Dr. Paul Broks
Clinical Neuropsychologist

I ASK MY STUDENTS IF THEY'VE HAD ANY THOUGHTS ABOUT what Frank's problem might be. I'm not expecting much but sometimes they surprise me. So, any thoughts? There's a long silence before a young woman, the talkative one, raises her hand. You can probably count Alzheimer's out, she says. Because? Because his recognition memory and his new learning and recall are better than you'd expect, at least for visual stuff. She makes the valid point that verbal memory is hard to assess because of Frank's word-finding problems, and the same goes for general verbal IQ. But his performance IQ, his nonverbal intelligence, appears to be at a good average level, which would be consistent with estimates of his pre-morbid ability. So, I summarize, as far as we can tell, there's no overall decline of memory or general intellectual function of the sort you might expect with a dementia. Silent assent. Right, so far, so good. Now what about the problem he has identifying parts of his body? Where does that take us? Someone else chips in, a guy who never says a word, and he's done his homework. It's called *autotopagnosia*, he says, and it's usually caused by problems in the parietal lobe. Which one? *Left*.

So, how do we read this? Well, I say, answering my own question, it's an indication that we might be dealing with something quite localized here, and you might be thinking of a tumor or a stroke, but there's nothing on the scans to indicate either of those, just a bit of cortical atrophy, not much, probably just age-related. So, it's shaping up as something that isn't a typical dementia, or a typical stroke or tumor, but it's causing serious problems for this man and it isn't getting any better. Present tense. Frank returns to life in these sessions.

I project a PowerPoint illustration of the brain's blood supply to indicate the location of Frank's problem, which is an arterial narrowing, a *stenosis*, in the left vertebrobasilar system at the back of the head, causing a relatively slight, but evidently significant, reduction in the blood supply to the posterior left hemisphere. It's not the sort of thing that would show up on a standard CT scan, which is all he's had up to now. The thing is, I explain, it's a condition which, unlike dementia, is potentially curable. They're all ears now. They've got to like Frank, and they're rooting for him. It's possible to treat this condition surgically with a procedure known as an endarterectomy, which is an operation to remove the stuff clogging up the artery so as to get the blood flowing normally again.

They want the denouement and, for a second, I'm tempted to give the Hollywood ending, but I never do. So, I explain, with someone Frank's age, seventy-five, it's not that straightforward. Seventy-five's quite old for this type of surgery. The surgeon has to be persuaded he's fit enough to undergo the operation and, obviously, Frank has to want to go through with it in the first place. Well, it turns out he is fit enough. He's ex-army and he's kept himself in good physical shape, but he's not keen on the idea at first. Dead against, in fact. I don't know why. He didn't have the words to explain. But, anyway, he's offered the operation and he gives it some thought, and decides he will go through with it after all. So, he goes through all the pre-op assessments and he's admitted to hospital, all set to go, but on the morning of the operation he changes his mind. He discharges himself at the very last minute. I still can't say why. Just didn't fancy it, I suppose.

I sense their disappointment. You know, I tell them, I have to admit I was pretty pissed off with him at the time. I had no right to be, it was his decision, but I was. Frank, you stubborn old bugger, I thought, *come on*! You've come through all these assessments— psychological, medical, four-vessel angiography and all—and you're nearly there. One last step, and it could really make all the difference. He was sat in a side room on the ward in his suit and tie with the regimental tie pin, ready to go home. I tried to persuade him, gently, professionally, keeping my frustration to myself, but he wouldn't budge. And I could see in his eyes, *just leave me alone, will you?* So home he went. And then his problems got worse and a few months later he had a massive stroke, and another one a few months after that, and that was it.

I gather my stuff and tell them, next week, we're going to be discussing an interesting case of anarchic hand syndrome. That's when a person's hand seems to act of its own accord, against their will, as if it has a mind of its own.

Snapshots

M Y FATHER, WHO WAS FOREVER HAMMERING AND FIXING things and making stuff, made himself a photographic enlarger. He took up photography after losing an eye in an industrial accident, an experience that left him with an acute sense of both the wonder and the fragility of vision. The camera, for him, was as much a tool for intensifying the process of looking as for recording images. Bathed in red light, I would stand and he would stoop in the cramped, makeshift darkroom under the stairs. I slid papers into the developing tray and was transfixed by the images appearing from nowhere: misty woods, spiders' webs, wet cobblestones, wrinkled old faces. I had some notion of the physics and chemistry of the process, and the technology, but it still felt like conjuring spirits, a fusion of magic and science.

My dad had a talent for portraiture and took a beautiful black-and-white photograph of Kate when she was in her early twenties, not long after we'd met. I can't find it. It's not where I thought it might be, among the other mounted pictures, the factories, the frozen ponds, the pipe-smoking policeman. I have, though, stumbled upon a batch of six-by-four snapshots. I tip them out of their shoebox and spread them across the carpet. There's a fat baby stuffing its face; two small boys in a tin tub; a boy standing next to a Christmas tree. Six months? Four years? Five or six? Ages are approximate. Most of the photos are undated. Moving on, there's a boy in football strip (Wolverhampton Wanderers) with the studs of his Top Dog football boots resting on a dubbined leather ball. He's about nine. Then there's a morose twelve-year-old eating sandwiches from a Wonderloaf wrapper; a thirteen-year-old in cricket sweater and striped shirt. The fifteen-year-old's pale, adolescent frame is half submerged in a black pond. The eighteen-year-old sips beer against a white wall in monochrome sunshine: long hair, Lennon specs. Then on to the next decade, and the next. There's a young man in his graduation gown. An older man wearing a suit. There's man and wife; man, wife and children. I know this unfolding life. It's mine. I know the story. But beyond name and genes, I'm not sure what connection I have with the protagonist. What sense does it make to say we are the same person, the middle-aged man that I am and the chubby infant staring out from the glossy gray distance of the 1950s? What strikes me is how little I remember. And what I do remember is often inconsequential: the furniture, the wallpaper.

Here's a case of a sudden and profound amnesia. Picture an elegant woman in her early forties. Let's call her Roberta. Stepping down from the train at her hometown station after a business trip to London, she suddenly becomes confused. She has the sense she's heading home, but why is she getting off at Exeter? Where is home anyway? And where has she been? The questions tumble rapidly, inexorably, to an endpoint: *Who am I?* She sits in the station café for a while to gather her wits, perplexed but outwardly calm. She retrieves a small mirror from the shoulder bag she's carrying and studies her face. It is her face, she knows that well enough, although it looks a little older and more careworn than expected. She finds a driver's license and reads, alongside the photograph, a name, evidently hers, a date of birth, and an Exeter home address. She rifles through the bag for more clues as to who she is, but there's not much to go on. She has presence of mind but no presence of self. The contact names in her phone mean nothing to her but, hesitantly, she keys the number listed under "Home." No reply, just the sound of her own voice inviting her to leave a message. How clear and confident and *knowing* she sounds.

It's a ten-minute taxi ride from Exeter St. David's station to the address on the driver's license. The large Edwardian terraced

house is unfamiliar. She rings the doorbell. She rings again. No answer, so she reaches for the keys in the bag and lets herself in. She
sits for an hour or more in the unfamiliar living room as the light
fades. Then a stranger turns up. He's not at all surprised to find her
in his house, but wonders why she's sitting in the dark.

Roberta's brain showed some borderline abnormalities of
blood flow in the frontal lobes, but beyond that the scans were
uninformative. There were no holes in her head to account for the
catastrophic drainage of memory. Her loss of identity suggested
a "psychogenic amnesia"—in other words, that the underlying
cause of the memory loss was psychological, rather than neurological, but nothing presented itself as a cause. She saw psychiatrists,
psychologists, hypnotists and therapists, to no avail. We pieced
together the story of her life—the schools, the jobs, the friends,
the colleagues, the relationships. She'd been married and divorced
by the age of thirty but had no recollection of her husband. "He's
a good-looking man," she said, studying a photograph of the pair
of them. They'd shared a flat for three years in Bristol. Had she no
recollection? "I recognize the lampshade," she said.

That's what I'm finding, going through the photographs. I'm
recognizing furniture and wallpaper more than I'm making connection with my child self. There's a clear sense of objects, space
and place but no sense of event. No sense of being there. I don't
remember my little brother washing my hair in the tub. I scarcely
even remember my first day at school, although there I am in
blazer and cap. I have no memory of lying like that in front of
the fire with notebook and biro. There are well-rehearsed, generic
memories of such things, long since woven into the fabric of "me,"
but the "I" of subjective experience is disconnected from the story.
Even when specific memories seep through I can't always be sure
they're reliable. The Swiss psychologist Jean Piaget remembered
being the victim of a kidnap attempt as a small child. He recalled
sitting in his baby carriage and witnessing the struggle between
his nurse and the would-be kidnapper. He saw the scratches on her

face and he saw the man being chased away by a policeman with a short cape and white baton. The event became a vivid paragraph in the Piaget life story, illustrated with rich visual imagery. Years later the nurse confessed to fabricating the entire episode. Did I witness the puppy dog being crushed under the wheels of a car, or is that someone else's story? I don't know.

Roberta and her two sisters spent an afternoon going over old photographs. How could she deny they were her sisters? There they were, together on the beach building sandcastles. There they were at Roberta's birthday party, nine candles on the cake. There they were, all smiles, nestled in their father's arms. How could she deny she was Roberta? She did not, but nor could she quite accept it. Three months into the amnesia her father fell ill and died. She had got to know the man and had grown quite fond of him, but she shed no tears at the funeral.

The phone rings. It's my brother, Pete. I tell him about the photographs. I remind him of the two of us sharing the tin tub. Then there's the picture of us sitting on the kitchen table playing banjo and accordion. Here we are on Bonfire Night trapped in rings of sparks. *Remember, remember.* There's the one of us in the river the day he nearly drowned. They illustrate my story, our story, but what does it mean to remember other than to recount the tale? "Not much," he agrees. We talk about the old house, a slum with holes in the roof, a history of suicides and reputed hauntings: 13 South Street. As well as buckets to catch rainwater, the attic was still littered with the previous occupant's religious paraphernalia and relics of exorcism. We were next door to a fish-and-chip shop. The lavatory was thirty yards away across someone else's backyard. Remember? Of course he does. It's part of us. If the street hadn't been demolished, I'd be back there excavating, looking for relics of myself in the bricks and tiles and corrugated iron fences. "I remember the carpet," Pete says.

St. Ives

WE CAME DOWN TO CORNWALL IN 1999 FOR THE TOTAL eclipse of the Sun. It was clouded out. Still, we sat on a beach and, at 11:11, the appointed time, the day went dark and the birds got agitated. Across the bay a lighthouse started flashing. A year later I took a job in Plymouth and we bought a house upriver, overlooking the wooded, western slopes of the Tamar valley. Cornwall was now our home. I loved the chill autumnal nights looking out across the valley with the stars churning overhead. The total eclipse was on the 11th of August, and it was three years later, to the day, that we discovered the lump.

One blazing midnight, out on the balcony, I'm picking out the constellations. Kate's not listening. *Cassiopeia, Pegasus, Pisces.* At least I thought she wasn't. *Perseus, Aquarius, Andromeda . . .* But now I'm hearing the story of the Chained Princess, Andromeda. Her mother, Cassiopeia, declared her more beautiful than the sea-nymphs and this had really pissed Poseidon off. So, as, mother watched from the shore with bitter remorse, the god of the sea chained Andromeda to a rock and left her for the ravenous sea-monster, Cetus. But a hero, Perseus, arrived in the nick of time to save her. He happened to be passing by on his way back from de-capitating Medusa the Gorgon. It was love at first sight. With the monster cutting through the waves, Perseus put the rescue on hold just long enough to observe the formality of asking Andromeda's parents for her hand in marriage, and then he slayed Cetus with his sickle or, in a variation of the story, by holding aloft Medusa's severed head, the sight of which turned the monster to stone.

"Cetus? There, below. And, over to the left . . ." I'm sharing

Kate's line of sight, directing her finger, as she had directed mine
to the tumor in her breast. "That's Perseus, charging to the rescue,
and that tiny smudge of light is the Andromeda galaxy." She can't
make it out. It's hard to fixate. The trick is to look fractionally to
one side so that the Andromedan photons strike the periphery of
the retina. That's where the cells most sensitive to dim light are
located. We've digressed from mythology to astronomy, and now
to anatomy, and soon she will lose interest, or, rather, patience.
It's all too remote; too removed from the worries of the world. The
reason for my love of the stars is precisely the reason for her indif-
ference. I'm enthralled by the notion that this smallest, faintest celestial speck of light, captured by the rods of the retina, my retina,
at the end of a two-and-a-half-million-year intergalactic journey,
and channeled for analysis along the visual pathways to a remote
fold of my brain, is an island universe containing a trillion stars.
I'm enthralled; staggered; enchanted. Kate isn't. For her it's a cold
fact. She doesn't get jazz either. So we sit in silent disjunction for
a while and then she says, "I'm dying. All this will go on without
me."

The days between the discovery of the lump and the diagnosis of cancer were dismal because I knew in my bones what the
diagnosis would be. A fog rolled up from the Tamar and wouldn't
shift. It was almost a relief when the bad news came and we could
get on with doing something about it. The sun was shining again.
A couple of days after the bad news we drove with the boys from
our home in the east of Cornwall to the seaside town of St. Ives in
the west. We sat watching the breakers crashing onto the sands
of Porthmeor beach. We swam, we ate fresh fish at the Seafood
Café, and we enjoyed the sun. It was a good time, a pocket of love
and warmth. There was a shared, unspoken feeling of whatever
happens, happens, but here, now, at this time, life is good. On the
first day I strode purposefully into the cold surf. I was clutching a
silver dolphin, a solid, well-crafted piece I'd been carrying around
for years, a lucky charm. I swam out until the figures on the beach

were minuscule and I flung the dolphin as far as I could out into the ocean.

THERE'S A SUDDEN gust of wind. The woman's wig flies off and bowls across the supermarket car park like a cartwheeling cat. The man gives chase. *Ha ha.* Funny. Except that I am the man and the woman is my wife, and she is crushed. She stands, rooted, hands on her baby-bald head, eyes welling. "Don't worry, Mum," our son reassures. "No one saw." Later, in the middle of the night, I lie suspended between resolve and despair. Kate is sleeping. Her skin smells faintly of chemotherapy.

WE RETURNED TO St. Ives the following year, after the surgery and the chemo, just Kate and me this time, and we went again to Porthmeor beach. We swam and, ludicrously, I kept an eye out for my silver dolphin, which, by now, and possibly for the next ten thousand years, would be buried in the seabed. After the swim I laid out a towel and settled to read a Ray Bradbury short story, "The Kilimanjaro Device." A man shows up at an old saloon in the wilds of Idaho and gets talking to the locals. He's a time traveler in search of Ernest Hemingway. When he finds him he will offer him an apposite death and a resting place in the snows of Kilimanjaro, beside the frozen carcass of a leopard, which is far better than his true fate. Let him die in the plane crash that almost took his life seven years earlier, rather than sink in a swamp of booze and depression. There are right and wrong times to die.

Kate recounted a childhood memory. I closed my eyes and listened. She's seven years old, on a day trip to the seaside with her mum and dad. She sees herself on the beach, standing at the water's edge. Another girl, about the same age, floats by, face down, lifeless, like driftwood. She recalled the commotion and then the quiet procession from the beach. No one was inclined to stay. The image of the dead girl surprised me. She bobbed in a flowing white garment, wreathed in seaweed like an infant Ophelia, watched over by swimsuited children with buckets and spades. The sea itself was subdued. Small waves broke indifferently. The whole scene appeared as if by projection, with no effort on my part. It was Kate's first encounter with the brute fact of death and, having relived it, she returned to her book and I returned to mine.

Later that day we went to a jeweler's shop and bought a wedding ring. Kate's fingers had swollen with the cancer treatments and the original no longer fit. She might have had it resized but we decided instead to buy a replacement, an unspoken reaffirmation of vows. "A year to the day," she said, and I saw us back in the consulting room getting the bad news from the oncologist and the Bad News Nurse. Everything was small, like looking through the wrong end of a telescope.

Dolphins

THERE ARE NUMEROUS DOLPHIN STORIES IN GREEK MY-
thology. We have already heard of Apollo taking the form
of a dolphin on his crossing from Crete to Delphi, to found his
temple on the slopes of Mount Parnassus, and later we shall see
how dolphins bore the hero Theseus down to Poseidon's palace at
the bottom of the sea in order to retrieve the ring that King Minos
had tossed overboard on their journey to Crete. Another story
involves Dionysus, the god of fertility and wine, who, returning
from his travels, was captured by Etruscan pirates mistaking him
for a wealthy prince they might hold to ransom. They tried to tie
him to the mast but the ropes fell apart on contact with his body.
The helmsman realized they were dealing with a god, not a mortal
prince, and urged the crewmen to free Dionysus and beg his forgiveness, but the sailors mocked him as a fool and raised the sail.
Yet, despite a good wind, the ship remained motionless. It started
to sprout vines and was quickly overgrown. The oars turned into
serpents and Dionysus transformed himself into a lion and chased
the terrified sailors around the ship. They jumped overboard to
save their skins and, as they leapt, turned into dolphins. For the
rest of their lives their job would be to help seafarers in peril. The
Greeks were great seafarers and dolphins were considered by sailors (who often sported dolphin tattoos) to be a good omen.

It wasn't only gods and demigods who had wondrous dealings
with dolphins. There's the story of Arion, the Dionysiac poet, renowned as a harper second to none, who journeyed to Italy and
Sicily where he earned vast sums of money through his sublime
skills as a musician. Having amassed great wealth, he decided to

return to Corinth and so hired a ship with a crew of Corinthians, whom he felt he could trust, a trust wholly misplaced because the crewmen were soon hatching a plot to steal his riches and throw him overboard. Arion got wind of this, but his pleas for mercy fell on deaf ears. The sailors gave him a cruel choice. He could either commit suicide right there and then, in which case his body would be returned to Corinth for a decent burial, or he could leap straightaway into the sea. He asked for a brief postponement of his fate so that he might stand on deck in his full minstrel's garb and sing a last song to life, at the conclusion of which he would put himself to death. This pleased the sailors, who knew they were in for a treat from the greatest of all musicians. So Arion donned his minstrel robes, took his lyre and sang, and when the song was finished he threw himself into the sea. The ship sailed on to Corinth leaving him for dead, but, as he plunged into the water, dolphins appeared and one took him on its back and bore him to the shores of Tainaron, from where he made his way to Corinth, well ahead of the treacherous sailors, who eventually met justice when their crime was exposed.

Dolphins were not just willing helpers of the gods and rescuers of mortal seafarers in distress. In mythology they were also psychopomps, creatures who mediate between life and death, escorting the souls of the newly deceased to the afterlife and acting as guides to the underworld. Especially privileged souls were borne by dolphins to the Isles of the Blessed, a paradise inhabited by heroes. Dolphins have been assigned a similar function, at least symbolically, in the modern era. In Carl Jung's depth psychology, which probes the unconscious workings of the mind, they have a psychopompic role in finding channels of communication between the conscious mind and the underworld of unconscious mental operations. More generally, they denote resurrection and metamorphosis.

The Disconsolations of Philosophy

SOMEWHERE ON THE OUTSKIRTS OF BIRMINGHAM IN THE INdustrial heartland of the English midlands, among the canals, the scrap heaps and the stockyards, there was a wall daubed with a big question: WHERE WILL YOU SPEND ETERNITY? When I was small, I looked out for it from the train. It troubled me. I knew what it was getting at. With adolescence came atheism, but, having banished the fear of eternal torment, what bothered me now was the prospect of eternal nothingness. It had always bothered me, I suppose, and I still see my child self confronting that smooth, gray wall, infinitely deep, infinitely tall, at the edge of the universe, stomach churning, trying to imagine the nothingness beyond. Death, you come to realize, is the only doorway to that unimaginable nonworld on the other side.

The idea of nonexistence is easy to grasp when applied to dinosaurs (extinct), unicorns (imaginary) and circular squares (logically impossible), but applying it to ourselves is unsettling. We were nothing before we were born, and each night we pass through virtual nothingness in dreamless sleep. Some have the misfortune to enter the void of prolonged coma. (*Where was I?* patients would ask, and the honest answer was, *Nowhere.*) We readily accept the science-fiction notion of being put into a state of suspended animation for an interstellar space flight, and there are people who make arrangements to have their body (or, at cheaper rates, just their severed head) frozen at the point of death in the hope of being revived and restored to conscious personhood when medical science has the wherewithal. Our minds can accommodate forms of temporary nonexistence, but never-ending nothingness, *absolute nothingness*, is another matter.

Sometimes I think of Kate in a timeless nowhere. It seems a bleak and lonely image at first, but, on reflection, it really isn't. It's nothing. In life, when someone goes away you can be with them in imagination, even if you don't know where they are. You know they are *somewhere*, and you can see them in your mind's eye doing something, anything: at work, walking down a street, driving, smiling, sleeping, breathing. But the dead, in their state of being dead, are beyond the reach of imagination. That's what you have to accept. People die and go out of existence. Indulge in wishful thinking of an afterlife if you will, but I think it's best to face that brutal, natural fact. So, I remind myself that Kate is nowhere now. That she was somewhere once, and the warm knowledge that she shared her life with me is all that matters. *What will survive of us is love.*

Arthur Schopenhauer, a towering figure in the landscape of early nineteenth-century philosophy, was disturbed by the thought of nonexistence:

> *A man finds himself, to his great astonishment, suddenly existing, after thousands and thousands of years of non-existence: he lives for a little while; and then again comes an equally long period when he must exist no more. The heart rebels against this, and feels that it cannot be true.*

Schopenhauer's vision of the world was bleak, uncompromising and, he knew, unwelcome. Elsewhere, he writes, "I shall be told . . . that my philosophy is comfortless—because I speak the truth; and people prefer to be assured that everything the Lord has made is good. Go to the priests, then, and leave philosophers in peace!"

Schopenhauer's summation of a human life as a sliver of consciousness inlaid between slabs of eternal nothingness is, surely, as true as it is stark. My heart rebels against it and feels that it cannot be true, but better to confront a harsh truth than swallow a consoling lie.

SCHOPENHAUER THOUGHT life was pointless. A human life is an infinitesimal flicker of consciousness in infinite space and time, and, within our puny lifespan, we are condemned to exist only in "the ever-passing present moment." Time is a destroyer, a vessel in whose wake all things pass into absolute nothingness. Real happenings, once they dissolve in the flow of time, have no greater claim to reality than events that never happened. "The whole foundation on which our existence rests," says Schopenhauer, "is the present." So to exist, bound to the restless, tumbling, present moment, is to be inherently unsettled. "We are like a man running downhill, who cannot keep on his legs unless he runs on . . . Unrest is the mark of existence."

We spend our lives striving for goals seldom attained, and even when we achieve them we are all too often disappointed. We are shipwrecked in the end, and enter the harbor stripped of masts and rigging. "Life was never anything more than a present moment always vanishing; and now it is over." By then it's all the same whether we've been happy or miserable, because life is behind us, all our days, all our joys and sorrows, utterly erased.

"Oh, for God's sake, cheer up!"

There's something vaguely familiar about the speaker. It's a fat man in a suit, fortyish, balding, the knot of his necktie halfway down his chest. He has a pint of beer in one hand, a cheap ballpoint pen in the other, and, staring out to the horizon, seems to be speaking to no one in particular. We're sitting, several tables apart, outside the Port William Inn on the north Cornish coast, facing the breeze of Trebarwith Bay. "Get a life!" he says, rising unsteadily to his feet and knocking over his now empty beer glass, which rolls over the edge of the table and smashes on the ground. "Get a life." He stumbles away and, as he goes, I notice he's wearing one of those hands-free earpieces.

I get another drink and watch the sun go down. It's what I used to do with Kate. We would drive to Tintagel and then head south along the coastal path here to Trebarwith Strand, and it was a walk we did all the more often with the first signs that the cancer was spreading. Three years post-diagnosis, Kate had small tumors in the spinal column and the hip, nothing incapacitating, but it was only going to get worse and there was a sharp sense of time running out. One walk in particular remains bright in my memory. We decided to drop by at the cliff-top-parish-church of St. Materiana. The churchyard was a picture of entropy, of a slow swallowing into the earth, ivy-clad tombs and headstones sinking hellward. I made a note of some of the inscriptions. Here's one:

> *Thomas Newbery Birch.*
> *Who passed into life eternal on All Souls' Day 1925. Aged 47,*
> "BEHOLD WE COUNT THEM HAPPY WHO ENDURE."

Christ could have had the angels lift him from the cross but the meaning of the crucifixion was *endurance*. The Christian promise: stick it out, keep the faith, and ye shall have eternal life. The great lie.

There was no sense of entropy inside the empty church. Time had stopped. In the north transept we found five votive candles set on a small table, just one of them lit. Should we light another one? I wondered. We didn't. I put a five-pound note into a collection box on the way out and we pressed on toward Trebarwith, into the winter sun, stone walls and bare fields to the left, the ocean to the right, three hundred feet below. The track turns close and closer still to the edge of the cliff, a hop and a skip from *life eternal*.

I felt the urge to leap. I always did. *Behold we count them happy who endure.*

Halfway along we sat on a slab of rock and stared out to sea, and caught sight of a pod of dolphins. No more than glints on the water at first, and even through binoculars it took a while for the patterns of light to cohere. There were half a dozen or so leaping and diving, threading water and air. We lost them when they reached Gull Rock, the black pyramid rising from the middle of the bay. I captured it in the circle of my binoculars but still could find no detail.

I look across to Gull Rock now and I feel a stab of absence. It's then I notice a sheet of paper flapping in the breeze on the fat man's table. It's weighted by a half-full beer glass. Now that's odd. It was empty. I saw it roll off the table and smash but there's no sign of broken glass on the flagstones, and I didn't see anybody sweeping up. I was miles away, obviously, out across the bay, halfway to Newfoundland. I can see there's a drawing on the piece of paper and I take a look. It's a diagram in the form of a rectangular box bisected by a partition. SPACE is written on one of the end sections. An arrow points left to right at the long base of the box, beneath which the word TIME. The word PAST is written to the left of the central partition, and FUTURE is written to the right. The partition itself bears the legend NOW and, below that, to the left, there's a little "x" with a connecting line looping out to large letters shouting: RIGHT HERE, RIGHT NOW!!

I look close at the palms of my hand, deep into the ridges of the skin. *I'm still here.* Right here, right now.

On Having No Head

"THE BEST DAY OF MY LIFE—MY REBIRTHDAY, SO TO SPEAK—was when I found I had no head..." These are the opening lines of Douglas Harding's philosophical memoir *On Having No Head—Zen and the Rediscovery of the Obvious*. Harding, an English architect-cum-mystic philosopher, is recalling a shaft of enlightenment experienced while out walking on a still, clear day in the Himalayas. He was thirty-three years old and had for several months been preoccupied with the question: *What am I?* But now he had stopped thinking. Words failed him. Past and future dropped away. "I forgot who and what I was ... there existed only the Now..." He stopped and looked, and this is what he saw: "... khaki trouser legs terminating downwards in a pair of brown shoes, khaki sleeves terminating sideways in a pair of pink hands, and a khaki shirtfront terminating upwards in—absolutely nothing whatever!" There was a hole where his head should have been, but it was no ordinary hole. It contained the world, and the real miracle, says Harding, was that the whole shimmering Himalayan scene, the snow-topped mountains and the blue sky mysteriously suspended in the void, "was utterly free of 'me.'"

Standing on Cornish clifftops with a head full of ocean and sky I am sometimes lifted by feelings of expansion and awe, of being almost absorbed into the scene. *Almost.* It's no more than a pale reflection of Harding's epiphany which was a magical awakening from the sleep of ordinary life, an unveiling of the perfectly obvious, "a ceasing to ignore something which (since early childhood at any rate) I had always been too busy or too clever to see ... what

had all along been staring me in the face—my utter facelessness."
The feelings of peace and joy were overwhelming.

Once, while dissecting a human brain, I was struck, viscerally, by another way of seeing the head, not as a space buzzing with mental life but, rather, as a chunk of meat and bone, the upper chamber of which is packed with a kilo and a half of brain stuff—the stuff weighing heavy in the palm of my hand. The average human head would tip the scales at around five kilos. Try loading five kilos of groceries into a shopping bag; let's say a liter container of milk, a half-kilo tub of butter, a kilo of carrots, the same of potatoes, a bottle of wine, a tin of baked beans and a bar of soap. That will give you an idea of the mass of the object you so effortlessly cart around atop your neck. From this perspective the luminiferous, universe-engulfing void above Douglas Harding's shoulders becomes a lump of matter, as much a part of the physical landscape as the mountains, rivers and rocks. The self disappears, but not into shimmering infinity. It renders down to the fats and fibers of biological material. But it just as surely disappears. Looked at this way, your head is as utterly free of "you" as Harding's snow-capped mountains and blue sky were free of him.

The Wrong Head

"GET THIS HEAD OFF ME," HE KEEPS SAYING. "GET IT OFF, IT'S the wrong one!" This is on the hospital ward. This is Jeff, an electrical engineer. His wife, Lucy, gives a shrug and a half smile of despair, eyes elsewhere. She is pretty, and he is handsome, or was. A red, raw surgical scar now tracks from the top of his shaven head toward the left ear. His face is relatively unmarked but a loosening of tone in the lower half makes him dribble. Eyes and forehead are locked in perpetual anger.

"Get this head off! It's the wrong one!"

Where do those thoughts go? Those memories? I am thinking of the car crash that has left him this way. Or, rather, I am thinking of the minutes before the collision, the innocent last moments of ordinary consciousness. It was the usual commute from work and around six o'clock, as usual, when he approached the bend in the road at the Chequers Inn. He would, I suppose, have been doing the usual things: listening to the news on the radio or playing a CD; sifting scenes from his day; watching the world scroll by; looking forward to an evening at home. He has no recollection of any of it, or of the days before. Where do they go, those thoughts and memories? They just go.

"Get it off!"

The driver of the hatchback that came skidding around the bend at the Chequers Inn, invading the cozy six o'clock interior of Jeff's car, and thereupon the interior of his head, was eighteen years old. He did not survive. You will still see a little shrine at the roadside, restored each week with fresh flowers.

"Get this fucking head off!"

"Shush, Jeffrey."

He's wrenching at his neck now. Lucy takes his hand and holds tight.

"It's the WRONG ONE!"

A few weeks later I follow up with a home visit. Jeffrey sits stiff and unspeaking at the kitchen table as his wife prepares coffee. The wrong head is resolutely still and quiet. Not in the mood. The original, the right one, housed the steady dispositions of a loving husband and father, the brain of an educated man. It was now a place where thoughts and urges roamed untethered. There were dark turns of mood and spiteful outpourings of abuse. Foul stuff, hurled mostly at Lucy. But she had a coping strategy. "When that happens I tell myself it's not really Jeff," she says. All the same, she stands by him out of the dutiful conviction that, at some level, deep, deep down, at the magical, essential core we all imagine inside others and ourselves, it *is* really Jeff. It is and it isn't. "You're a bit low at the moment, aren't you, love?" she says, but he's beyond melancholy.

We are handed mugs of coffee and Lucy leaves us to it. For the next hour, sitting in the kitchen with the refrigerator buzzing, I enter the void of Jeff's frontal lobes, probing with questions and puzzles and neuropsychological exercises. "Tell me as many words as you can think of beginning with the letter 'A.'" "Ant," he says, after a while.

I worked for a time in a psychiatric clinic and found depressed people hard to deal with. They might feel a photon brighter for their hour of cognitive behavioral therapy but I'd leave the session soaked to the bones with their despair. Depression has an almost physical presence. With Jeff it's different. It's an absence that fills the room. He is a black hole. The longer I spend in his company the emptier I feel. I watch his coffee going cold, untouched. I watch a filament of saliva break from his lip. *Entropy, entropy . . .* My thoughts dissolve. We sit in silence.

I lay out some picture cards between us on the Formica

breakfast bar. They're jumbled up, I explain, but you can rear-range them to tell a little story. Go ahead. See if you can make a story. He looks at me, he looks at the picture cards, and then again at me. Slowly, deliberately, still holding my gaze, he brushes the cards one by one off the edge of the table, on to the floor. He's had enough, and so have I. I'm packing my things away when a small child enters the room. Jeff sweeps him up into his arms and kisses his cheek. "My son," he says. It isn't his son, as a matter of fact, it's the neighbors' kid, but I don't let on.

I drive home, taking a detour via the Chequers Inn. There are more fresh flowers at the shrine. Where do they go, those thoughts and memories? They just go.

Apophany

THE MAN WITH THREE NIPPLES HAULED THE BODY FROM the pool. It lay limp and unresponsive despite the pumping and pounding. "How long was he under?" someone asked. Thankfully, I didn't know; it wasn't my shift. I'd only just arrived. We gave up, but then the old man spewed another gallon of water and wheezed back to life. According to the report, his name was Milligrew.

I spent several summers working as a lifeguard. That year it was a glassy new municipal pool in the West Midlands. As well as the man with three nipples, I was working alongside an international powerlifting champion and an escaped convict. A typical midmorning snack for the powerlifter would be a whole roast chicken and two pints of milk. We didn't know our colleague was a criminal until the police came to take him away. He was good company and I was sorry to see him go. His replacement was a waster by the name of Pusey, who didn't last long. One day the leisure services director turned up with a local dignitary for a tour of the site and they discovered our friend in the plant room sharing a joint with a shivering nymph in a bikini.

Three years later I was lifeguarding in Oxford, sometimes at Hinksey Pool, sometimes at one or other of the riverside bathing places, more often than not at the murky backwater known as Tumbling Bay. It was hidden behind the allotments at the edge of Botley Park. No one ever went there. One day I'm sitting in my wooden hut, out of the rain, reading a newspaper, and I chance upon the name Pusey. It's not the same person, but a train of memories is set in motion: the man with three nipples, the near-corpse, the powerlifter, the armed robber and louche, lazy Pusey. I wonder what he's doing now. That same evening I drop into the Corn Dolly pub, which is not one of my usual haunts. There's a punk band playing. They're terrible and my brain, primed by beer and reminiscence, starts playing tricks, because each time I look at the bass player I find I'm hallucinating Pusey. Later, he materializes right next to me at the bar and I realize that it is in fact Pusey. The very same. It seems oddly inevitable. "Amazing coincidence," I tell him. "You hadn't crossed my mind in years until this afternoon. And now . . ." "Yeah," he says, unimpressed. I'm not sure he even remembers who I am. Then he's off and I never see him again.

The Pusey coincidence spooked me. Carl Jung, who coined the term "synchronicity," believed that "at least part of the psyche is not subject to the laws of space and time." He steeped himself in quantum theory and the *I Ching* and came up with the *acausal connecting principle*. Psychological phenomena are a fundamental constituent of the universe, he believed, and the world is threaded through with patterns of connectivity governed by the meaning of events. Meaningful coincidence—synchronicity—is an expression of this. "Oh come," said Sigmund Freud on one occasion. "That is sheer bosh." I am inclined, for once, to agree with Freud. But I can't shake off the eerie feeling.

Perception is all about attaching meaning to patterns. Our survival depends both on the ability to identify regularities in the world and to respond to irregularities. Generally, we know what to expect. We have an intuitive sense of cause and effect, and of what is probable and what is improbable. In the tumultuous cascade of people and events that constitutes a lifetime, random improbable conjunctions are bound to occur. That they are down to chance doesn't make them any less disconcerting. We still feel impelled to seek an explanation, some superordinate pattern.

All creative endeavors are geared toward the discovery of new patterns and connections, fresh ways of seeing the world, but sometimes the struggle for insight can be counterproductive. Epiphany (a moment of realization) has an obscure cousin in the lexicon of madness: apophany. It refers to the point at which an ordinary experience becomes the fountainhead of delusion. The newsreader says, "Good evening," and you know at once that he is Satan himself. Your neighbor's car catches the light and you realize that he and his fellow extraterrestrials are bent on destroying your brain with their deadly rays. People vary in their susceptibility to seeing connections between seemingly unrelated events and ideas and, to paraphrase Seneca, all imagination has a dash of madness. But weird coincidences can induce a psychotic wobble in the sanest of minds. You get that vertiginous sense of alienation from consensus reality, that there are more things in Heaven and Earth.

A couple of days after my encounter with Pusey I arrived for my afternoon shift at Hinksey Pool. I was taking over from Nick, the mountaineer, and Hazel, the wall-of-death rider's assistant. Hazel was writing up an incident report. An old man had slipped, cracked his head and fallen, dazed, into the deep end. Nick had fished him out, Hazel had patched him up. He seemed to recover well enough and they'd sent him on his way. The man's name was Milligrew.

The Cave of the Skull

I N BOOK VII OF *THE REPUBLIC* PLATO IMAGINES PRISONERS IN A cave, shackled so as to lock their view on to the wall directly in front of them. They've been there since childhood and know nothing of the goings-on behind them, let alone of the world outside the cave. Above and behind them, at a distance, a fire is blazing, and between them and the fire is a walkway with a low wall. People cross the walkway carrying all manner of objects—vessels, statues, figures of animals—and their shadows are cast on to the wall in front of the prisoners, who take them for reality. The shadows, for them, are the truth. A prisoner released from his shackles and forced to confront the world around him would be dazzled by the glare of the fire and confused by the strange, solid forms of the passersby, who would seem far more alien and illusory than the shapes of the shadow world.

Scholars have different interpretations of Plato's cave allegory. Some see it as primarily a representation of *the problem of knowledge*. How (if at all) is it possible to get a true reflection of reality in the distorting mirror of the human mind? Others see it more as a depiction of a mindset of human ignorance, maintained for political ends by the powers that be. Arguably, both interpretations are valid.

I rather like the image of the cave itself—the blazing fire, the flickering shadows, the shackled prisoners oblivious to the traffic behind them, and the endless passing carnival of statue-bearers, themselves oblivious to the prisoners. I like to think of the cave in literal terms. We are prisoners of biology. We see the world and ourselves confined within the cave of the skull.

The Rich Thicket of Reality

V ISUAL ILLUSIONS PROVIDE A CLEAR DEMONSTRATION THAT
the world we experience is a fabrication of the brain. We are
all prey to the perceptual deception of illusions such as the "Table-
top Illusion," devised by the American psychologist Roger Shepard.

The tabletops are identical in size and shape, yet, because of
its orientation, the one on the left automatically triggers a stronger depth-perception response in the visual system and so appears
elongated. The brain is misapplying rules of perceptual adjustment
that would normally serve it well in its task of world building. Illusions are intriguing and tantalizing. They undermine the intuition that *seeing is believing* (it's as much the other way about) and
create an irresolvable conflict between what we experience and
what we know. Those tabletops look manifestly different, and it's
a misperception that's impossible to override with knowledge of
the "true" state of affairs. We know it to be true that the tabletops
are identical but we feel it to be false. The mismatch between the
world of physics, in which the tabletops are measurably the same,
and the *world of phenomena*, in which they are different, confirms
that the latter is a (sometimes faulty) product of the brain/mind
factory. *World of phenomena* is just a fancy way of referring to the

plain, ordinary, everyday world of objects, people and events as we automatically and effortlessly experience it; the world we take for granted as the *real world out there*; the world as it appears to be. In the case of the tabletops there is a puzzling conflict between *the world as it appears to be* (the tables are different) and *the world as it is* (the tables are the same). The "truth" of the matter is settled by use of a scientific instrument—a ruler—but they still *look* different, which is another kind of truth. Visual illusions give a glimpse into the machinery behind the brain's construction of the phenomenal world (colors, shapes, objects). Their deeper significance is in revealing that *all* of our percepts, reliable and otherwise, are constructed by the brain, not just illusory ones. The psychologist Chris Frith puts this nicely when he says that our perception of the world is a fantasy that coincides with reality.

The distinction between reality and imagination is also an aspect of brain function. Let me simply define reality here as what goes on without us after we die. Imagination dies with us. While we are alive, reality and imagination interweave, but, in general, it's a good thing for the brain to maintain a clear distinction. Psychosis is one consequence of a collapse of the partition between real and imagined, inner and outer. The institutionalized insanity of religious fundamentalism is another. But the capacity to hold rational thoughts alongside irrational intuitions is part of the mind's design. Only modern Western culture has tried to eradicate magic and the demons of the supernatural realm, creatures of an Otherworld taken for granted by all other cultures since the dawn of history. We should take this long psychological heritage into account if we want to understand our own minds.

Even if we deny belief in the supernatural, we are all inclined toward magical thinking and superstition. It's a frame of mind that in one direction opens out to a dream world of myth and imagination and in another leads to practical creativity in the arts and sciences. In the words of William James, we stand "in the rich thicket of reality" yet sense there is something beyond the visible world, "a more spiritual universe," which gives our lives significance. Most of our imaginings, serve practical, problem-solving purposes. We pause to think, *What if I were to do this rather than that*? We picture ourselves in the counterfactual universe of *this* or *that* course of action. We examine the different consequences and, sometimes, the more bizarre the imaginings the greater the insight. Sixteen-year-old Albert Einstein imagined pursuing a beam of light, a fantasy that led ultimately to the special theory of relativity. *Imagine a world without thought experiments*, the joke goes.

To the modern mind magic is unbelievable, but I like to toy with the idea that the mind, or at least that aspect of the mind that we call consciousness, is itself unbelievable, as unbelievable as the gods and ghosts of the Otherworld. But being unbelievable is not to say it isn't *real*. If we are to get anywhere close to understanding consciousness, the first thing to acknowledge is its absurdity.

Hello, Anybody There?

ROLAND SOMETIMES SLIPPED INTO A PARALLEL UNIVERSE. It was a world in which he had once served as a bodyguard to Princess Diana and where, in the course of his duties, he had stumbled upon a terrible secret. Consequently, undercover agents were now pursuing him and he feared for his life. I asked him what the terrible secret was. He closed one eye and pinned me with the other. "The wolf child," he said.

Diana had given birth to a baby girl covered head to toe in animal fur. The doctors knew something was wrong from the early stages of the pregnancy and the princess had been advised, then urged, and then ordered, to have a termination. She refused, saying she would love and care for the child come what may. She was, after all, a good, loving person. Everyone knew that. But when it came to it she couldn't stomach the horrible reality and her husband, Prince Charles, had had the monster dispatched. Roland heard voices, too, which he located inches above and to the right of his head. "About there," he said, circling the empty space with his cigarette. For the most part it was a semicoherent mumbling that didn't trouble him too much, but sometimes they came through with loud and abusive clarity. They'd told him they'd get him sooner or later.

You couldn't challenge the Diana story. It was unarguably true for Roland and interrogation only made him angry. But the voices were another matter. He agreed there was something wrong and was ready with a neurological explanation. Part of his brain had become functionally detached from the rest (this was the work of the secret agents) and was now semiautonomous, issuing statements and streams of thought that sometimes mingled with the main flow and sometimes just gurgled alongside. "I'm neurologically possessed," he said.

Neurologically possessed. The juxtaposition of science and the supernatural was unsettling. "Possession" evokes medieval images of demons and dark forces. It doesn't figure in the lexicon of modern psychiatry. Still, I thought the way he used the term captured something of the essence of psychosis, the sense of invasion and yielding of control, the dispossession of thoughts and actions: the merger of *self* and *not self.* For most of us the solitariness of private consciousness is also sanctuary, a space in which thoughts and actions, desirable or undesirable, are securely one's own. Roland's inner sanctum of consciousness had been violated. The words and thoughts in his head were not exclusively his. It must have felt like possession and, at the neurological level, perhaps in a sense it was. Auditory hallucinations are usually troublesome, sometimes terrifying, and generally taken to signify mental illness. But voice-hearing might also be considered a part of normal experience. There has been debate over the pathological status of hallucinations ever since Jean-Étienne-Dominique Esquirol first gave them a clinical definition in the middle of the nineteenth century. Around one in fifty people admit to hearing voices from time to time and the bereaved are especially prone. A friend tells me her dead husband often speaks to her. Sometimes they're words of comfort; at other times he makes her laugh as he did in life. She doesn't believe in ghosts. The voice, she thinks, is testament to a relationship so long and loving that one mind has braided with another in the circuits of a single brain.

I, too, have heard voices, just a handful of times, and fleetingly. Arriving in New York once from London, I stepped off the plane and straight into a long afternoon of meetings followed by an evening of social events. It was the early hours of the following day before I finally made it back to my hotel room. I fell instantly into a deep sleep, and then into a lucid dream in which I was holding my baby son, who was wrapped in a multicolored silk scarf. I could smell his skin. The colors and textures of the scarf were hyper-real. I was wide awake by six and the room was filling with a gray light. The first voice, just to my right, put a question: *Is that it?* The response came from the left: *Probably.* There was a rush of fear, not because I thought there were intruders in the room but because I knew for sure I was alone. The voices resembled a typical schizo-phrenic hallucination in some ways. They sounded real and "out there," rather than inside my head, but in other respects they were not typical of psychosis. There was a simple question and response; four bland words that did not address me directly or give a running commentary, as psychotic hallucinations probably would. The ex-perience would qualify as a "hypnopompic" hallucination, which is a false perception on awakening, so can be safely filed away as "normal." I was terrified all the same. It was more than the sense of losing control. My multisensory "hallucination" of baby and scarf was beyond my control but it was joyful rather than fearful. Fear of the voices was more to do with the deep strangeness of the feeling, however illusory, that there were other articulate beings cohabiting my head. It did not amount to "neurological posses-sion" in Roland's terms but, certainly, there was a passing sense of having been invaded.

That was five years ago and it didn't happen again until the early hours of this morning. As with the New York experience, today's episode followed a period of sleep deprivation and was as-sociated with a sequence of vivid dreams: being on a broad, stone bridge in dense fog with horsemen looming fast from the gloom; a large band of cyclists pedaling furiously and singing in beauti-ful harmony; a motorcyclist crashing fatally into a bridge but-tress; the interior of a building with white stairways and corridors blocked by white walls; a male voice calling unintelligibly from somewhere behind one of the walls. Next, I'm in bed with a young woman. We lie in silence for a while, close together, half-clothed, and then without a word she leaves, at which point I'm woken by what sounds like the chiming of a clock, but there's no clock in the room. As I lie awake trying to figure out if it was a real sound or a dream sound, there's another signal, something like a smartphone alert. But my phone is switched off. Then I hear a voice, which sounds like mine. *Hello?* it says, in a tentative, enquiring sort of way, *Hello, anybody there?*

This time around I'm not the least bit afraid. I want to know what else the voice has to say. Perhaps it's my Socratic demon trying to get through. Socrates, one of the founders of Western philosophy, the advocate of critical self-analysis whose method of reasoning through rigorous argument has helped shape the character of the modern mind, was a voice-hearer. His friend Xenophon recollects him saying, "a divine voice comes to me and communicates what I must do." In Plato's account, Socrates says that his *daimonion* "always dissuades me from what I am proposing to do, and never urges me on." In other words, it intervenes when he is about to make a bad decision. He considered it a form of divine madness, a gift from the gods akin to the gifts of poetry and love. As it happens, there's been something on my mind this past couple of days, an emotional matter, with a straightforward binary choice to be made. I must take one of two paths. I'm still at the crossroads pondering and maybe my demon already knows what's for the best, but he can't quite get through. *Hello, anybody there?*

Sleep and dreams: our nightly dose of death and psychosis. What was that one all about? Putting Freudian interpretations to one side, and considering, literally, the architecture of the dream, the bridge and the labyrinthine white building suggest nothing so much as the architecture of the brain. It's a journey, via the corpus callosum, from one hemisphere to the other. Perhaps, in line with Roland's theory, a part of my brain, somewhere down those white corridors and stairways, was functionally detached from the rest.

Is that it? Probably.

Bessie's Diary

THE ERIC RAVILIOUS PRINT HANGING ON THE WALL OF MY study, *Dr. Faustus Conjuring Mephistophilis*, was a leaving gift from a colleague at a hospital I used to work at. It's a woodcut. Faustus, the polymath-sorcerer, kneels at the center of a pentacle, clutching his heart. A naked Mephistophilis, the Devil's emissary, springs through a black doorway. The good angel and the evil angel clash in the shadows. Faustus is a master of philosophy, medicine and law, but, frustrated with the limitations of traditional forms of scholarship, he has pledged his soul to Lucifer in exchange for magical powers, which, he thought, would bring fame and wealth as well as infinite knowledge. Magic has failed him. It has diluted his reason and blurred the boundaries of reality and imagination. I feel the picture is doing the same to me, just a little, imagining a dark energy flowing between it and a small black diary lying on my desk, Bessie's diary. I have an impulse to move the diary out of view. Bessie, I am fairly sure, would have found the picture disturbing. No, worse, I have a sense that she is, right now, feeling wary of it.

Bessie Renaut was Kate's great-aunt. Her little black book fits the palm of my hand. The spine is crumbling but the pages are sturdy and only a little yellowed. She wrote in flowing copperplate with scarcely a correction or crossing out. "Diary. China 1900," it begins. "June 23rd." Later that year the *Missionary Herald* noted Bessie's characteristic self-deprecation in offering her services. "I know God can and does use the weakest instruments for His work," she wrote. Then it went on to describe her martyrdom at the hands of the Boxers, the rebellious anti-imperialist, anti-Christian, Fists of Righteous Harmony. They had their own mission, which was to eradicate foreign devils.

Having fled to the hills around Hsu-chou, Bessie and her small band of foreign devils were discovered and briefly imprisoned, before being dragged by their heels through the streets of the city, after which the veins in their arms were slit and they were left to die. That was the 9th of August. She was twenty-nine years old. I'm not sure how the diary survived, but it found its way to Kate's grandfather, Bessie's not-yet-born cousin, and now it rests on my desk, solemn as a tiny coffin. I feel a little wary of Bessie's book, sensing her presence, and her disapproval. I won't yield to superstition but, still, I feel uneasy. The little black book has the aura of a sacred relic. I have a typed transcription, but it's not the same. The book is imbued with Bessie. The chemistry of her conscious brain was transduced to marks on the page by her own steady hand. Molecules of skin and sweat—her physical self—were impressed into the paper as she wrote. Her anxiety is there, too, toward the end, in the more urgent slope of the script: urgent but, to the last, restrained. She never abandoned her religious formulations: "We take comfort from God's word and cast our care on Him who careth for us." Some caring, it seems to me, for God to visit agony, humiliation and brutal slaughter on those who spread his word. But so be it. He moves in mysterious ways.

Why do I feel uneasy? Angels and devils, as depicted in the engraving, are figments of the medieval imagination. I believe that nothing remains of Bessie beyond her tragic story. Immersion in the rationalism of clinical neuroscience for thirty years left me with no reason to believe in immaterial souls. There is no ghost in the machine, and if we did find one, then we'd only have to start looking for the machine in the ghost. Why, then, should I feel there's something ghostly about a book? In personifying Bessie's diary, my guard has dropped and left me vulnerable to an irksome obsessional thought, soaked in primitive magic, which would be neutralized by the compulsive act of putting the diary out of sight in a drawer. Again, I resist. Instead, I open the book.

> *July 21st—Suddenly 2:45 an attack. This was made from ground above, grit, stones and boulders being hurled at the mouth of the cave by about fifty or sixty men. The gentlemen went out and fired . . . Mr. Parker shot at a man who persistently hurled at Mr. McKellen.*

Bessie and her fellows were charged with the same religious zeal: they spent weeks living together in caves, hiding in holes, half-starved, and yet do not appear to have been on first-name terms. The Boxers scatter. Mr. Parker examines the wounded man, a young captain, and takes from him his charms and native medicines. "We are going to wash his wounds," writes Bessie, "and if too weak to be sent off, drag him up so as to be able to keep the wolves off him." I wonder what became of the man. There is no further mention. Did he return with his band to drag the foreign devils to their deaths?

The Great God Pan

PHEIDIPPIDES (C.530–490 BCE) WAS A LEGENDARY GREEK
runner who, having just fought in a grueling battle, ran non-
stop from Marathon to Athens with news that the Athenian army
had defeated the invading Persians. "Joy!" he cried. "We won."
Then he dropped dead. This may or may not have happened. The
story does not appear in the writings of Herodotus, the "Father of
History," and the prime historical source for the Battle of Marathon. But another Pheidippides story does. With the Persians
coming into shore at Marathon he was charged with the task of
running from Athens to Sparta with a request for military support to repel the invasion. He covered the 240-kilometer distance
within two days, across rough terrain and in the blistering heat
of August. The Spartans were sympathetic to the plight of the
Athenians but unable to help immediately because they were in
the middle of a sacred festival. Only when the Moon was full, and
the festival over, could they march to Marathon. That was a week
away.

Pheidippides set out immediately on his return to Athens with
the dispiriting news. Along the way, high on Mount Parthenion,
he heard the call of his name and was confronted with the horned,
cloven-hooved figure of Pan, the god of the wilderness, of shep-
herds and flocks, whose very appearance could inspire raw terror. Now, though, the goat-god's mood was benign. He seemed
puzzled. Why was it that the Athenians paid him no attention he
wondered. Why did they not honor him? After all, he was well dis-
posed to them, had been of service on many occasions, and would
be again, soon. Yet still they ignored him. Ask them, he instructed.

Pan's message, as well as the Spartans', was delivered, and then, from Athens, Pheidippides forged on to Marathon to join his comrades in battle with the Persians. It was a battle which, against the odds, and without the support of the Spartans (who arrived too late for the action), they won. Pan, legend has it, was instrumental in the Athenian victory, making his appearance on the battlefield and disabling the Persian forces with waves of extreme, irrational fear. *Panikos!* The Athenians duly built a shrine to him at the foot of the Acropolis and held an annual ceremony in his honor with sacrifices and a torch-race.

What are we to make of all this? The classical scholar Philippe Borgeaud writes of Pheidippides' encounter with Pan as "only a projection of his wish," a consequence of the "tension, depression, and exhaustion" caused by constant running. It's an entirely rational explanation, but has been challenged. Charles Boer, the literary scholar, thought we should take Pheidippides at his word and understand Pan "as the splendid imaginal reality he was." Prior to the fifth century BCE, he contended, the people of Greece were quite used to the intervention of such beings. They could see them, hear them and touch them. They were not, "at least in their eyes," making them up.

These two views of Pan, on the one hand an exhaustion-induced apparition and, on the other, a solid figure of "imaginal reality," are not incompatible. You could say they were both forms of hallucination, but perhaps we're missing something if we lump them together and dismiss them as *mere* hallucination. Perhaps there is something more particular to that expression, *imaginal reality*, than is contained in the generic clinical term *hallucination*.

We usually place imagination and reality in opposition. The talking animals, centaurs and fauns of C. S. Lewis's land of Narnia are imaginary creatures inhabiting an imaginary world. The people, bears and wolves of Siberia are real inhabitants of a real place. It's straightforward, surely. Just as a physical object can't be in two places at once, something is either real or it's imaginary. The inhabitants of Narnia can't pay a visit to Siberia, or vice versa. But let's not be too quick to dismiss "imaginal reality." Perhaps there's something to it. Am I suggesting that the Great God Pan could materialize in front of me, as real as the FedEx man who knocked at my door this morning to deliver a parcel? Might I one day open the door to be confronted by a minor Greek deity with horns and cloven hooves? Yes and no. "No" because Greek gods don't really, truly, objectively exist, *out there in the real world*. Of course they don't. Then again, it's possible that the ancient Greeks' experience of the gods went well beyond conventional products of the modern imagination, and given that their brains were, structurally, identical to yours and mine, it's perfectly conceivable that I *could* open the door and find standing before me, in broad daylight, a creature I recognize as Pan. There's a knock at the door and there he is with his hooves, his horns, his fur and, slightly worryingly, his large, erect penis. He speaks, explaining that he is in pursuit of a fugitive nymph, and wonders if perhaps I have offered her sanctuary. Pan is inordinately fond of nymphs, although, if he's in the mood, a shepherd boy will do, or even a sheep. If none of the above, he's fine with solitary gratification. He did, after all, invent masturbation. I do a reality check. I pinch myself, good and proper. He looks real enough. I can see in fine, bright detail the color and texture of his fur. His shiny horns catch the sunlight. His deep, gruff voice sounds real enough, too. He speaks excellent English with a Greek accent. Would the accent of a speaker of ancient Greek resemble a modern Greek accent? Evidently, yes.

"Well?" he says.

I'm shaking with fear.

"No. No nymph here. Definitely not."

Pan looks suspiciously over my shoulder and sniffs. Dare I touch him to see if he's real? I dare not. But, with a clack of hooves, he steps forward and grabs my wrist and there's no doubting his physicality. I'm close to panic now; stricken with the fear of Pan. "You sure?" he says, dropping his big-chinned face close to mine.

He smells goatish. "I'm absolutely sure." Perhaps he's an ordinary mortal man in fancy dress. No. Those hindquarters are unmistakably goat-jointed. I'm about to faint with fear but then he grunts and loosens his grip. He turns in a flash, and is over the hedge in a single bound. I pinch myself hard. It hurts.

Suppose I'm telling you this now as a true story, as something that really happened. What do you make of it? There was a knock at the door and I had an edgy conversation with a minor Greek deity as real as the FedEx man. Honest. I saw him clearly. I heard him. I felt him and I smelt him. I'm a fantasist. I'm insane. These things have already crossed your mind. It can't really have happened because it's a bedrock fact that Greek gods don't really, truly, objectively exist *out there in the real world*. Do they? Well, perhaps they do. Perhaps it depends what you mean by *really* exist. You're edging away. You don't want to get into a dope-head debate about the nature of reality. But consider the possibility that gods and, for that matter, ghouls and goblins, have an authenticity, a variety of existence, which may not have the full-blown, measurable, weighable reality of, say, a door or a dog but which lies partway in that direction, somewhere between doors, dogs and daydreams, a kind of partial reality. I am going to play Devil's advocate and try to persuade you of the existence of this realm of gods and ghouls, without for a moment dropping my naturalistic, scientific guard.

I am going to defend the view that the gods of ancient Greece were indeed sometimes perceived by (some) people as physically present, real beings and were not, in the usual sense, figments of imagination, and that they were something beyond our conventional understanding of "hallucination." This will, I'm afraid, require some consideration of the nature of reality (so light yourself a spliff), but the central strands of my argument will be drawn from the well-documented experiences of people who enter an altered state of consciousness known as "awareness during sleep paralysis," *sleep paralysis* for short, a realm betwixt reality and dreams.

Something Wicked This Way Comes

"*C*ARLA!*"

Carla hears her name in the dream but that's not what wakes her. What wakes her is the cat padding on her shoulder. It must have got in through the window. She tries to turn but can't. Her limbs are dead meat. She is paralyzed. Only her eyes move and they find a stranger standing in the doorway. Her heart heaves, but the man just looks incuriously across the room and goes on his way. At his feet is a small bay horse, the size of a dog. They are just the first of the intruders. Now the room hums with electrical energy and a gnomish old man appears. At his side, waist-high, squats a creature, half-spider, half-crab. The old man scuttles forward. He leans close over Carla and spits hard onto her eyelid. She feels the impact and the wetness. And then she feels his fingers at her throat.

Carla is locked in a state of sleep paralysis. Time can slow almost to a standstill. That's what worried her friend Joe, another young sufferer. It might only be thirty seconds, he said, but it can feel like hours. This was a cause of deep concern for Joe. If the mind can stretch seconds to hours, why not days? And if days, he thinks, why not years?"

"*Carla.*"

It's a sad, soft, gentle voice now, her father's. He leans over and kisses her forehead. A tender gesture, but Carla is trapped, can't respond, can't move. Not a finger or a toe, not a millimeter. Carla is dead in the casket. Now she is in the earth. She feels the pressure all around. There is no air to breathe.

"And if years, why not eternity?"

—∞∞∞—

WE'RE SWAPPING SLEEP paralysis stories over pints of stout in a busy London pub, but I can't compete with Carla. My sleep paralysis episodes are tame by comparison. Following the typical progression, they usually start with a buzzing *whoosh* in the ears, and then it's paralysis and panic. I just *know* there's something evil lurking in the shadows and I'm terrified, despite all efforts to cling to a rational perspective. It's sleep paralysis! It's harmless! I picture the brain activity: the increase in blood flow to the non-rational right hemisphere; the amygdala (the brain's threat detector) going into overdrive. It makes no difference. But, for me, that's as far as it goes. My experiences don't progress to the terrifying apparitions of witches, goblins and ghouls. And I've had maybe five or six episodes in my entire life. Carla MacKinnon, an intelligent and level-headed thirty-two-year-old, sometimes gets five or six a week. She's a filmmaker who, in the process of researching a short film about the condition that torments her, has gathered the testimonies of a remarkable group of fellow sufferers, some of whose experiences are yet more extreme than her own.

Sleep paralysis is a symptom of narcolepsy, a rare neurological disorder that causes excessive daytime sleepiness, but most sufferers, like Carla, are not narcoleptic. Nor are they mentally ill, a common fear. Around one in three people will experience sleep paralysis at some time in their life. It can be triggered by irregular sleep patterns and excessive alcohol. For a minority, the paralysis, the panic, the evil presence (and sometimes a crushing sensation) are merely the prelude to a yet more horrifying phase in which the firewall between fantasy and reality collapses and all hell breaks loose. It's as if some chamber of the collective unconscious has been unlocked, releasing all the monstrous archetypes: witches, vampires, goblins, demons and all manner of other strange creatures.

Sleep paralysis is a wellspring of myth, from the night demons of ancient Sumeria to the sex-crazed incubi and succubi of medieval folklore and the Night Hag of Newfoundland. Native South American mythology has the shape-shifting Amazonian river dolphin, or *boto*, which at night takes human form to prey on women in their beds. He has to wear a hat to hide his blowhole, which remains in situ despite the shape-shift. And the ancient Greeks had Ephialtes, another shape-shifter who took different forms to menace his victims but who was most commonly identified with Pan, the libidinous god of the wilderness, whose horns and hooves became the Devil's trademark in medieval Europe. Those space-age fiends of the night, the extraterrestrial "Grays" of alien-abduction lore, are the latest incarnation of the evil night intruder.

Sleep paralysis is a strange and disturbing state to fall into, but let's pause to reflect upon the strangeness of ordinary sleep. If you sleep an average of eight hours a night and live for eighty years you will have slept a solid twenty-six years over the course of your lifetime, and for around a quarter of that time you are dreaming, which is to say you will have spent around six and a half years wandering through a world of involuntary images and disjointed logic. Each night we retire to our beds and lose contact with reality, and ourselves, until the morning when our brains once more draw the threads of self and world together, setting us up for another day of ordinary consciousness in the ordinary world. It's quite weird when you stop to think about it. The discovery of the underlying patterns of brain activity in some ways deepened the mystery of sleep.

Sleep science came of age in the 1950s with the work of two Chicago physiologists, Eugene Aserinsky and Nathaniel Kleitman. Their primary tools of investigation were the electroencephalogram (EEG), which records the electrical activity of the brain via scalp electrodes, and the electrooculogram (EOG), which measures eye movement. Using these methods in combination, they discovered that sleep has an elaborate, hidden architecture.

Far from being a simple switching-off, a typical night's sleep is a journey through well-defined cycles of brain activity, each with a patterned progression of EEG brainwave changes. But it was the eye-movement recordings that were crucially revealing. It was common knowledge that sleepers' eyes will sometimes dart back and forth behind the lids. What the Chicago researchers discovered was that these eye movements were associated with a particular, and surprising, pattern of brain activity, and also with dreams. On this basis they divided sleep into two types. REM sleep (for "rapid eye movement") is a phase in which the eyes and the brain are highly active. It usually kicks in around ninety minutes after sleep onset. Remarkably, the rapid, low amplitude, electrical brainwaves associated with REM (beta waves) are similar to the patterns associated with alert wakefulness. For this reason REM sleep is sometimes referred to as "paradoxical' sleep. While to all appearances the person is asleep, their brain seems to be wide-awake. Prior to the REM phase, the sleeper has been through three phases of progressively deepening sleep, each with a distinct EEG signature but with no associated eye movement. This was designated NREM (for "Non-REM"). Dreams may occur during NREM sleep but they are less vivid and memorable, have less story content and lack the surreal quality of many REM dreams. Another important difference between the two types of sleep is that REM is accompanied by inhibition of the skeletal muscles, so-called *atonia*, which renders the person, in effect, paralyzed. By contrast, in NREM the voluntary muscles are perfectly functional, and it's during the deep, slow-wave phase of NREM that sleep-talking, sleepwalking and children's bedwetting occur.

Ordinary waking consciousness is richly textured and, with senses alert and tuned, strongly oriented to the external environment. With skeletal muscles fully functional, we are able to act in the world according to our perceptions and intentions. In REM sleep, with senses muted, consciousness is focused inward; we are not aware of the external environment, and the skeletal muscles are inhibited. With sleep paralysis there seems to be an overlap of wakefulness and REM sleep such that consciousness is, at one and the same time, sensitive to the internal and the external environments, dreamlike and real, while the skeletal muscles, as per typical REM states, are inhibited. So, on this account, Carla's brain gets jammed between sleep and wakefulness. At one level, fully conscious, she sees her bedroom and its contents in clear detail, but other parts of her brain are locked in dream sleep. She is trapped and helpless. The muscle atonia of REM sleep prevents us from acting out our dreams. In sleep paralysis it becomes a straitjacket. The apparitions of sleep paralysis have been described as dreams projected into real surroundings, which may be apt, but in some ways they are quite undreamlike. Ordinary dreams are predominantly visual, whereas sleep paralysis hallucinations engage all the senses. Carla felt the old man's fingers at her throat. She heard his foul whispers and smelled his breath. Dreams mostly involve interactions with fellow humans, but it's fiends and folklore characters that populate the world of sleep paralysis.

Nineteenth- and twentieth-century literature contains numerous descriptions of sleep paralysis-like experiences, including in the works of Thomas Hardy, Herman Melville, F. Scott Fitzgerald and Ernest Hemingway. Guy de Maupassant's short story "Le Horla," published in 1887, is a well-known example. This is the tale of a wealthy man in the process of losing his mind after impulsively saluting a beautiful, pure white sailing vessel passing by on the River Seine. This innocent act, he comes to believe, served as an invitation to a malign supernatural being to haunt his house and his mind. He is soon having nightmares of an invisible creature approaching him as he sleeps and trying to strangle him, "... getting onto my bed . . . kneeling on my chest . . . taking my neck between his hands and squeezing it." The creature, it seems, is the forerunner of a horde of extraterrestrials intent on annihilating the human race. Maupassant was well versed in psychology and neurology, having attended the lectures of the eminent neurologist Jean-Martin Charcot at the Salpêtrière hospital in Paris, and his psychological acumen was much admired by Nietzsche. But the depiction of sleep paralysis in "Le Horla" is probably based as much on personal experience as academic knowledge. With advancing syphilis, Maupassant was increasingly prone to sleep disturbance, hallucinations and paranoid thoughts. This tale of a disintegrating mind was told with the author's mind itself in the process of disintegration.

There's an almost textbook description of sleep paralysis in Bram Stoker's *Dracula*. Here's Mina Harker reporting her first encounter with the vampire:

> *There was in the room the same thin white mist that I had before discovered . . . I felt the same vague terror which had come to me before and the same sense of some presence . . . Then indeed, my heart sank within me: Beside the bed, as if he had stepped out of the mist—or rather as if the mist had turned into his figure . . . stood a tall, thin man, all in black . . . I would have screamed out, only that I was paralyzed.*

Mina's progression through fear and sensed presence to a menacing physical form coalescing from shadows or mist is characteristic of many sleep paralysis reports.

A staple of vampire lore, actually created by Stoker, is the idea that vampires are not reflected in mirrors. It was a discovery made by Mina's husband, Jonathan, during his stay at Count Dracula's castle in Transylvania:

> *I had hung my shaving glass by the window, and was just beginning to shave. Suddenly I felt a hand on my shoulder . . . I started, for it amazed me that I had not seen him, since the reflection of the glass covered the whole room behind me . . . there could be no*

error, for the man was close to me, and I could see him over my
shoulder. But there was no reflection in the mirror!

Through Carla, I came across a sleep paralysis case that immediately struck me as a true-life counterpart of Stoker's gothic imaginings, a neat fusion of Jonathan and Mina Harker's fictional experiences. Francine, a thirty-year-old software developer, had been diagnosed with narcolepsy in her early twenties. One weekend, staying over at a friend's house, she experienced an especially horrifying episode of sleep paralysis that came to a crescendo with the appearance of a wild-haired, fiery-eyed witch. The witch materialized from the shadows, moved swiftly to the foot of the bed and leapt upon her in a frenzied attack, spitting and cursing and digging sharp nails into her neck. There happened to be a large mirror propped against the wall at the side of the bed and, at the height of the attack, Francine turned her eyes sideward. Glancing in the mirror, she saw the room in the half-light of dawn, and she saw herself lying motionless on the bed. But nothing else. All was still. No sign of the witch. On one level the observation is mundane, merely confirming the fact that, however terrifyingly real they seem, visual hallucinations are not part of the objective, physical world. They are subjective experiences. And yet I can't help but feel a frisson of the uncanny when I picture the scene, with fury in the room and stillness in the mirror. It's as if, seeing through Francine's eyes, a part of me is persuaded of the reality of the monster. But what if the witch had been reflected in the mirror, I wonder, would that be so surprising? If our brains have the capacity to conjure up vividly realistic multisensory hallucinations (the creatures are "as real as real" one sleep paralysis sufferer told me), you might think they could go the extra step of conjuring up reflections in a mirror. On this evidence they can't. Mirrors, like cameras, don't lie, and the one propped against the wall in Francine's bedroom was a portal to reality. Alice's looking-glass in reverse.

Perhaps, in the way it fuses fantasy and reality, sleep paralysis

opens a window on to an ancient, but now largely hidden, dimension of human experience: the imaginal reality at the source of all
myth-making. *Imaginal reality* is a term I first came across in James
Hillman's book *Pan and the Nightmare*, which includes a lengthy
quotation from an article by Charles Boer, in which he argues
that, up to the fifth century BCE, the Greeks experienced their
gods as visible, audible and palpable beings. They were not mere
personifications of abstract ideas. Thus, Pan, when he appeared to
Pheidippides on Mount Parthenion, truly did appear, and truly did
give audible voice to his concerns about the Athenians, as might
a flesh-and-blood human being. Hillman claims that divine and
demonic figures were for millennia perceived as having a palpably
real presence, "But the scientific *Weltanschauung* with its cut between observer and observed severed us from that witness, and its
testimony became magical thinking, primitive belief, superstition,
insanity." I don't think anyone is suggesting that Pan had a physical presence such that you could hoist him on the scales and weigh
him. But the idea that Pheidippides' experience of him went beyond the bounds of modern imagination, and perhaps even beyond
our current understanding of the word "hallucination," is worth
considering.

To pick up on a definition I tossed into the ring a little earlier,
let's accept the distinction between reality and imagination and
say (over-simply, but I think undeniably) that reality is *what goes on
without us when we die*, whereas our imaginings die with us. Now,
you can carve reality up in all sorts of ways—physically, socially,
conceptually, etc.—but it is, essentially, independent of individual
minds, and in that sense objective rather than subjective, exterior
rather than interior. The world will go on without us when you and
I are gone. It would go on, at least physically, if all life on Earth
were suddenly to be obliterated in a cosmic catastrophe. Fantasy,
by contrast, including imagery and counterfactual ("what if")
thinking, is interior and subjective. We may share the products
of our imagination to some extent through language and the arts,

but the source of imaginative experience is inviolably private. I am at this very moment conjuring in my mind's eye an absurd image. I can safely say that no one in the history of the world has ever pictured such a scene, nor ever will. I am not going to say what it is. It will remain a secret, and when I die the memory of the absurd image will die with me (if it hasn't already faded with brain-rot). That's all I mean by saying that acts of imagination are inviolably private and internal. The world will go on rolling around the Sun but my little flash of absurd imagination will have been and gone, unknown and unknowable to anyone else.

I can, in a manner of speaking, project my voluntary acts of imagination into the here and now, real world. I might, for example, picture Kate standing beside me. I can imagine having a conversation with her. But it's a pitifully thin representation of what the reality was. I can't, literally, feel the warmth of her hand, or hear the cadences of her voice. And she can say nothing in any imagined conversation that would truly surprise, amuse or inform me, because it is I who am generating the words. Reality and imagination are counterposed in this scenario but remain categorically distinct. The room I am sitting in has the familiar, real-world quality of exteriority, and the mental image of my late wife does not. Despite my strongest efforts to bring the details more sharply to life, it remains forlornly muted by comparison with the shapes and colors of objects in the room.

Imaginal reality, as I am trying to frame it, is a fusion of the imaginative and the real, not merely a juxtaposition. It's a partial breaking down of the partition between interior and exterior and so, to some extent, between subjective and objective. Imagery and thought processes are projected, involuntarily, into the real-world scene with full vivacity and thus experienced as exterior to the self. In imaginal reality, rather than being a ghostly-thin internal image, Kate would stand beside me as real as real. I would feel the warmth of her hand and hear the cadences of her voice. She might even tell me things I didn't know or hadn't thought of.

Placing these various forms of experience along the dimensions of autonomy (voluntary/involuntary), vividness (realistic/muted) and location (internal/external), we can differentiate imagination, ordinary dreams and imaginal reality as follows. The products of imagination are voluntary, muted and internal. For example, Carla imagines being greeted by a friend's dog. She conjures an image of the dog with its tail wagging, but the image is pale by comparison with the sight of a real-life, tail-wagging dog. There's no question that she has willed it into existence and that it is experienced as being located in the arena of her own conscious mind. Dream imagery, by contrast, is involuntary, and somewhat more vivid and external-feeling, such that we seem to enter into, and become enveloped by, a different world. Carla dreams that a mad dog is chasing her down the street. She didn't deliberately construct this situation in the way that she willed the images of the encounter with the friendly dog. She had no say in being dropped into the dream scenario, and it feels to some degree real, at least real enough for her to feel fear. And yet, the experience lacks the full-blown, multisensory, 3D quality of being chased by a mad dog in real life. Nor, come to think of it, is the setting quite right. She turned a corner and somehow found herself in the street where she lived as a child. It feels unreal. Dreams do. They are *dreamlike*.

The apparitions of sleep paralysis are vivid in their rich texture and multisensory detail ("real as real") and externally located in a real setting that can be scanned and scrutinized as the monsters go about their evil business. Carla sees, hears and smells the gnomish old intruder. She feels the wetness of his spit and the pressure of his fingers at her throat. The monsters have autonomy in the sense that they appear to speak and act in purposeful ways beyond the control of their victims, but then so do the less real-seeming characters of ordinary dreams. It's the killing combination of multisensory vividness, autonomy and appearance in a real setting that makes them so compellingly "real." But although the experience has all the qualities of wakeful consciousness, the apparitions depart from reality because they are, despite the compelling sensory evidence, subjective experiences. No one else can see them. They don't reflect in mirrors. They don't go on without you when you die.

Sleep paralysis hallucinations are convincing and terrifying enough, but I am going to call this the "weak version" of imaginal reality and, pushing the idea further in the direction of the Greek gods, will speculate that there might also be a "strong version." In the weak version, the apparitions are experienced as external to the self and have a terrifying degree of apparent autonomy in the sense that they are beyond the person's voluntary control, but they are mere hallucinations—mindless, subpersonal projections of the unconscious mind. They have autonomy insofar as they spring up of their own accord and occur alongside real perceptions (to paraphrase the definition of "hallucination" offered by the philosopher and psychopathologist Karl Jaspers) but, for all that, they are mindless and totally subjective. In the strong version of imaginal reality they go beyond this limited, mindless sense of autonomy. In the strong version they have some semblance of mind. They are in some ways extrapersonal (that is, independent of the person's mind) rather than subpersonal, and so have a degree of objective presence. As sleep paralysis sufferers will tell you, these creatures have the shine of sentience in their eyes, the look of intelligence and malicious intent, but such reports, however compelling the perceptions on which they are based, provide no evidence at all that the monsters have purposeful minds of their own. And I'm not suggesting that they do. *Quite.* (That would be the "very strong version.") What I am suggesting is that for a time in ancient Greece there may have been a kind of halfway house between the subjective and the objective, that there was something about Greek minds, and Greek culture, during that period in history that made people susceptible to multisensory hallucinations of archetypal figures—the gods—whose function was to influence a person's behavior through advice, guidance and warning. And to that extent the gods had a degree of independent personhood.

If there's any mileage in all this (and I know it's a big speculative leap) then sleep paralysis and its archetypal apparitions might stand as a model of the god-saturated psychology of the Greek mind. We moderns take it for granted that each of us has a unique, private, inner world of thought, an internal space where memories, thoughts, images and plans play out: the arena of consciousness. We might assume that human minds have always been structured this way. But, according to the classicist Bruno Snell, this is a relatively new way of thinking that began with the Greeks. In his influential work, *The Discovery of the Mind in Greek Philosophy and Literature*, Snell traces the gradual discovery of interior mental life, from the epic poems of Homer, composed around the eighth century BCE, to the writings of Plato some four centuries later. Cognitive advances through the middle third of the first millennium BCE, he argues, amounted to a revolution of the mind, a realization that we have a unique and individual inner world of thought quite distinct from the outer world (incidentally, thus sowing the seeds of the modern "mind–body problem," as we will see in a later discussion of consciousness). Our science, our literature and our philosophy all have their origins in that period of ancient Greek civilization. But the Greeks did not merely define new methods of inquiry, and topics for investigation, they created the very idea of human beings as individual, introspective, self-reflective, intellectual beings. As Hillman puts it, "Even the idea of an idea is Greek." Prior to this time, the boundaries between inner and outer, subjective and objective, self and other, would have been more fluid. You might be less certain that thoughts, utterances and intentions were your own because the very idea and sense of "your own" was less robust.

I can only guess as to whether this, in itself, would have made people more prone to intrusive, multisensory hallucinations, fantasy fusing with reality. But, when they did present themselves, apparitions of the gods (appearance, character and utterances shaped by clear cultural expectations) might not only have had the perceived reality that sleep paralysis creatures have, but also, beyond that, even a degree of agency, which is to say coherent thoughts and intentions independent of the hallucinator's. Such thoughts and intentions, filtered through the prevailing culture, would be fashioned by the hallucinating individual's brain but through systems operating independently of those responsible for maintaining his (relatively porous) ego boundaries, his subjective "I," and therefore not claimed as his own.

I suppose what I'm suggesting is little more than a variation on the Julian Jaynes theme of psychotic voice-hearing as a vestige of the "bicameral minds" of the Homeric heroes. In his much-disputed (detractors say crackpot) account of the history of consciousness, the American psychologist claimed that it is only recently—since the second millennium BCE—that human beings have felt themselves to be the authors of their own thoughts and actions. Prior to that, behavior was directed by hallucinated voices perceived to be of external, supernatural origin (although actually arising from the right hemisphere of the brain). "There is in general no consciousness in the *Iliad*," says Jaynes in support of his thesis. "And in general therefore, no words for consciousness or mental acts." Achilles and Agamemnon were, effectively, automata—puppets of the gods.

According to Jaynes, the bicameral (two-chambered) mind of our proto-conscious ancestors reflected a greater functional separation of the brain's hemispheres. That we are now autonomous and introspective is due not to changes in the brain's hardware but rather to software developments wrought by social and cultural change. But modern vestiges of premodern mentality abound. They are evident, for example, in the universal appetite for religious authority and ritual. Hypnotism and the auditory hallucinations associated with schizophrenia are other examples cited by Jaynes. Some of his psycho-archaeology may be suspect, but there is, in fact, good evidence that in some cases the voices heard by psychotic people are due to the misattribution of subvocalized speech. The brain's executive systems fail to monitor certain components of the speech production process and the inner voice takes on a life of its own.

You don't have to accept all of Jaynes's speculations to be stimulated by the central, bold idea that major historical shifts in brain function have driven the evolution of self-awareness. He's certainly had his critics, but Jaynes's broad theoretical vision has also attracted some prominent admirers among philosophers and cognitive scientists. Daniel Dennett, for one, takes the Jaynesian notion of "software archaeology" very seriously. But, whatever the merits of the bicameral mind hypothesis, I would suggest that the multisensory apparitions of sleep paralysis in fact take us closer to the image of the gods as they may have presented themselves to ancient Greek minds than does Jaynes's "divided brain" theory of hallucinations. I'm really not sure how helpful divided brain ideas are in this regard. As far as brain function is concerned, I've noted the possible REM sleep/wakefulness overlap hypothetically linked to sleep paralysis, but that's as far as I'm going.

A Grief Observed

How many hours are there in a mile? Is yellow square or round?
Probably half the questions we ask—half our great theological
and metaphysical problems—are like that.

—C. S. Lewis, *A Grief Observed*

I READ *A GRIEF OBSERVED* IN TWO SITTINGS, EITHER SIDE OF lunch, on a rainy Saturday. C. S. Lewis's reflections on bereavement were first published in 1961 under the pseudonym N. W. Clerk. Lewis had lost his wife, the American poet and novelist Joy Davidman, the previous year. He refers to her throughout as "H," her first name being Helen. Clive Staples Lewis disliked his own given names and from early childhood was known as "Jacks" or "Jack." He survived Joy by little more than three years, succumbing to a heart attack on 22 November 1963, the day that John F. Kennedy was assassinated. *A Grief Observed* was subsequently published under his own name, although Joy remains "H." Unknowingly, I purchased the book on the forty-ninth anniversary of the author's death, somehow less significant than the fiftieth might have felt, but a pleasing coincidence all the same.

From the first pages it was clear that my experience of bereavement was quite different from Lewis's, so much so that I began to question whether I had grieved at all and wondered what he would make of it. I nodded off with those thoughts circling my brain. It had been a late night and I'd had a glass of beer with my lunch.

I was woken by a tap-tap-tap at the window. The room was now semidark. "Why don't you ask him?" I opened my eyes to see a man standing right outside the window. He looked vaguely familiar, though I couldn't really place him: fortyish, plump, balding, crumpled gray suit, necktie low slung, a can of lager in his hand. He tapped and spoke again. "I said, why don't you ask him?" I got off the sofa and opened the window. "Why don't I ask him?" The man squeezed his fat face into an odd expression, arching an eyebrow. "Well," he said, "you know what to do."

Yes, I did. It was obvious. So I closed the window and went up to the bedroom. I ran my fingers down the lacquered walnut of the big old wardrobe we'd carted around with us for thirty years and four house moves, then I opened the door and stepped inside, in among the shirts and jackets and dresses. Strange. I thought I'd given all of Kate's stuff to the charity shop. There were fur coats, too. She never wore fur. I took a step further, one arm stretched forward, expecting to feel the woodwork at the back, but there was nothing. Of course there wasn't. This is how it works. So on I went through the pitch dark. There was something crunching under my feet. That would be the snow, wouldn't it? It was all very queer, but familiar at the same time. And then there was light, and I found myself standing under a lamppost in the middle of a wood on a snowy night with not a soul in sight. I hear a pitter-patter of feet coming toward me. It's a strange little person with the upper body of a man and the legs of a goat. He's holding a snow-covered umbrella over his head, and trots right by me, not stopping. I know I'm supposed to follow. The sun rises as we make our way. The snow melts. Winter turns to summer. We are out of the wood now. The little faun has darted into a bush and disappeared and I am left walking down a familiar street toward the University Parks.

It's a summer's day in Oxford. We're in the Parks sitting in the shade of a white willow that overhangs the slow-flowing waters of the Cherwell. The date is indeterminate. The man sitting next to me is lost in contemplation. Picture a burly, baggy-eyed man, early sixties, stooped and balding, overdressed for the weather in tweeds. He hasn't spoken for a long half minute but I know precisely what his next words will be. Here they come: "No one ever told me that grief felt so like fear." It's a rich, radio voice, though softer than the one he used on the wireless. "Fear" is a drawn-out conflation of "fair" and "far." His gaze stays fixed on the river.

"Fear?" I say. "No, not for me."

"Not fear as such," he says, "but something resembling fear; a fluttering in the stomach."

"No. Not even that."

There are punts gliding by. There are strollers and picnickers and lovers lolling in the grass, but still the date won't settle. The abundant feathered hair, the skinny boys in stonewashed jeans and jammers, and the brown-legged girls in denim cut-offs all signal the eighties, but some belong in quite another era, young Jack's possibly. He locks inquisitorial eyes on to mine and says, "You loved your wife?" And I say, "Yes. Very much."

I decline a cigarette. He lights one up. I tell him there were many moments of fear before she died. They came with a sudden weightlessness in the gut like driving too fast over a small bridge. There were routine flutterings of dread in hospital waiting rooms; missed beats in the middle of the night, feeling her warmth beside me, knowing that before long I'd be sleeping alone. But afterwards, when she was gone, there was not a trace of anxiety.

"If you loved her," Jack says, "you must have suffered."

"Yes. But you spoke of agony, and it wasn't that."

A punt goes by. The punter looks something like me. His girl-friend, trailing a hand in the water, could be Kate. They are soon under Rainbow Bridge and out of view. I'm enjoying the cigarette smoke mingling with the scent of new-mown grass. There's a damselfly hovering and darting in purposeful zigzags not far away. It's getting closer and Jack is uneasy. The creature stops dead still, pinned to the air, right in front of us, a little blue strip light. Jack tenses. I tell him they don't sting or bite, but that doesn't reassure him so, after I've taken a few seconds to study his subdued panic, a man who knew the terror of the trenches now in frozen fear of a damselfly, I brush the insect away. He's had a morbid dread of insects since childhood. It's their jerky, angular limbs and the dry, metallic sounds they make.

"To my eye," Jack says, "they are either machines that have come to life or life degenerating into mechanism." The bad dreams of his childhood were mostly of specters or insects. "And to this day, I would rather meet a ghost than a tarantula."

Joy Davidman entered Lewis's life in January 1950 when, through an introduction from the American writer Chad Walsh, she began a correspondence. Previously a hardboiled atheist and member of the Communist Party of America, she had been converted to Christianity some four years earlier, in significant part due to the influence of the writings of C. S. Lewis. She came to England in September 1952 and they met for lunch in Oxford. Joy was accompanied by her London pen-pal, Phyllis Williams, with whom she was staying. George Sayer, Jack's former pupil and future biographer, completed a foursome. Sayer described the occasion as "a decided success." Joy, he recalls, "was of medium height, with a good figure, dark hair, and rather sharp features. She was an amusingly abrasive New Yorker, and Jack was delighted by her bluntness and her anti-American views." She was thirty-seven years old; Jack was fifty-three.

With his dread of the damselfly subsiding, Jack returns to grief. "Agony, yes. Agony is the word. But there were moments when something inside me tried to persuade me that I didn't mind so much, not so very much, after all, and that I would get over it. I had, as they say, plenty of 'resources.'" He punches his right hand

softly into the palm of the left. "People do get over these things!"
Another soft punch. "There's more to a man's life than love!" And
another. "I'd been happy in the years before I met Joy!"

"But trying to be rational didn't help?"

"No. All that 'common sense' just made me feel ashamed. And
then would come a jab of red-hot memory and 'common sense'
would vanish like an ant in the flames of a furnace. I'd be reduced
to tears, maudlin tears, and it disgusted me. I think I almost preferred the agony. It was at least pure and honest."

"And the tears were not?"

"They were self-indulgent; self-pitying."

"There's nothing as pointless as self-pity. We can agree on
that."

Jack flicks the stub of his cigarette into the river and lights another.

"Your 'jabs of red-hot memory,'" I ask. "Did you have in mind
the boxer's punch or the hypodermic needle?"

"The punch."

"My word for it was 'stab.' Stabs of absence; stabs to the gut, to
the brain and heart; an entering of the flesh, a *knowing* in the flesh
that she's just not here anymore."

He shows no change of expression but my words have made a
mark because for an instant I see desolation in his eyes, but then
he draws on his cigarette and turns toward me, his generous features now shaped to curiosity.

"Where are you from? I can't place your accent."

"The twenty-first century."

JACK HAS SUGGESTED a bite to eat, so we're heading out on to
South Parks Road. The date still won't settle but mostly it's the
eighties. A Ford Sierra goes by followed by a Renault 14 and a succession of other vehicles that fit with the decade, but then along
comes an Austin 7 and for every ten people kitted out for the days

of Margaret Thatcher there's someone dressed like Lloyd George. I recognize a young woman walking toward us, richly hennaed hair curling down onto a white denim jacket with a CND patch.

"Katrina! Hey!" "Hi," she replies, having no idea who I am, and I can't explain. "Sorry," I say, "You don't remember me. Never mind. It was a while ago." It was for me, anyway. I want to say something warm and heartfelt and uncreepy, but can't think of anything. So I just come out with, "You're looking gorgeous," which in those days was a risky thing for a middle-aged stranger to say to a woman like Katrina, but she smiles and says, "Who are you?"

"You'll never guess."

I give a half wave and she gives a half smile.

"Pretty girl," Jack says.

"Long gone. Motorcycle accident."

The temperature drops rapidly. Clouds gather. Leaves are turning gold and falling from the trees. By the time we reach the end of the road the branches are bare. Giant snowflakes are sweeping down from impenetrable darkness above the streetlamps. Still in my jeans and T-shirt, I'm glad to get to the warmth of the Eagle and Child. Jack, in tweeds, is better attired for the change of season. We're in the Rabbit Room at the back of the pub, on to our second pint of ale and tucking into sausage and mash.

"And then there's the laziness of grief," Jack says. "At times I couldn't summon the effort even to shave."

"It's not laziness. It's lethargy. The fatigue of depression."

"'Depression' doesn't do it justice. The word's too . . ."

"Clinical?"

"*Finite.*"

"I couldn't say. I wasn't depressed. Nor was I numb or angry or any of those things you're supposed to feel."

"At all?"

"There were overwhelming waves of sadness. There were

tears and, by the way, unlike you I felt no shame in them, or self-loathing. But I wasn't depressed. The opposite in a way."

"The *opposite*?" He is shocked.

"I found reserves of energy and resolve. My head was clear. I felt strong."

"No despair, no confusion?"

"No."

"No thrashing like a drowning man?"

He leans back against the oak paneling of the alcove, retrieves a pipe from one pocket and a tin of Gold Block from another. We are soon enveloped in a swirling, aromatic fog. It's so thick I lose him for a few seconds.

Then he says, "You think I'm a washed-up wreck, don't you? A hopeless case."

"Not at all."

Four hard puffs on the pipe as he draws his thoughts together.

"The thing is, you know, it's the completeness, the finality of it, that so appalls."

"I know. It's hard to comprehend. But you, of course, believe in an afterlife, whereas I don't. I believe I will never see Kate again but you at least have hopes of being reunited with Joy."

"No," he says, "not in the sense of a happy reunion on the further shore. But . . ."

Puff, puff, puff.

"In any case . . ."

Puff, puff.

"The further shore! Those simpleminded, earthly images! I would willingly accept eternal separation from Joy, I'd accept being eternally forgotten by her, if these were the conditions for her joy in Heaven. But, I asked myself, Lord, are these your terms? I can meet Joy again only if I learn to love *you* so much that I don't care whether I meet her or not?"

What does he mean? There's too much faith, too much theology for me to get my head around the words. But then he recasts his thoughts in plainer terms, and I have a sense of what he's getting at. "I think of it this way. If I could have cured her cancer by never seeing her again, then so be it. I would have arranged never to see her again. Of course I would, and so would any decent person. Wouldn't you?"

"Yes. I don't know. I wonder. Any decent person? Is it really so straightforward?" I have a dozen questions lining up. But Jack has put another match to the bowl of his pipe, *puff, puff, puff,* and Jack fades in billowing clouds of blue tobacco smoke, and, when the smoke clears, he is gone. In his place, also puffing on a pipe, sits the fat man, the man who came tapping at my window.

I'D ENJOYED LEWIS's company. He was a Christian and I'm an atheist, and, to put it bluntly, I think when he most needed comfort and reassurance his Christian faith let him down. The shock of Joy's death undermined his beliefs. Through tortuous theological logic (it seems to me), he eventually restored them, but, for a time at least, his religiosity made matters worse, sharpening his sorrow. His God had let him down. I didn't have a god. I hadn't been let down by anyone. Something bad had happened in a quiet corner of a vast, indifferent universe. Nobody to blame.

There are many Christians, and other religious people, who gain succour from their faith at times of personal crisis, and of course there are atheists whose lives are shattered. So I'm not for one moment saying *atheism good, religion bad* when it comes to coping with bereavement. That would be unforgivably smug, as well as plain wrong. What interests me about Lewis is the peculiar convergence of intellect and imagination, and his tenacity in clinging to religious faith at any cost to the rational mind. He wrote of the poet W. B. Yeats that, "To put it quite plainly, he believed seriously in Magic," but also that he was a learned, responsible writer whose ideas were worthy of consideration. He might have been describing himself. He certainly believed in magic, as far as miracles are magic, and, for Lewis, miracles were part of a wider set of supernatural beliefs, including belief in ghosts, Heaven and the afterlife. The more I read—the fiction, the autobiography and the Christian apologetics—the more I came to appreciate that he was a serious thinker as well as an artful writer, and I began to wonder how it was possible for a refined, and otherwise rational mind to become so steeped in magical beliefs. The other thing that had intrigued me, a small but possibly significant detail of his life, was Lewis's fear of insects: *either machines that have come to life or life degenerating into mechanism.* The supernaturalism and the distaste for mechanistic little beasts, I thought, might be two sides of the same coin.

It was Lewis's close friend, Owen Barfield, who compared insects with French locomotives, both having "all the works on the outside." That was the problem, thought Lewis, *the works. The mechanism.* He recalled a nursery-book picture of a Tom Thumb figure perched on a toadstool and being threatened by a stag beetle much larger than himself. That was terrifying enough, but what made matters worse was that the beetle's horns were strips of cardboard separate from the plate and fixed to a pivot so that they could be opened and closed like pincers. To young Jack's sensitive mind the mechanism was an abomination. Whatever its origin, the phobia took root in the battleground of logic and magic, or, let's say, *mechanism* and *mentality.*

For Lewis the damselfly was a sinister threat, a miniature robotic menace. To the extent that the creature acts with purpose—to capture prey and avoid predators and other capricious hazards along the riverbank—it has *mind.* Second by second its tiny brain selects courses of action from a set of alternatives, but the damselfly's repertoire of behaviors is determined solely by its genes. It does not have anything like the reflective self-awareness and freedom to choose that we humans have. It cannot decide to turn vegetarian, eating moss and lichen rather than aphids and gnats. No one knows for sure but, quite possibly, the hovering, darting, bug-munching frame of the damselfly is entirely unconscious, an exquisite, purposeful but totally insentient biological robot. A machine "coming to life" or "life degenerating into mechanism." It blurs boundaries, and some people need definitive boundaries, especially when it comes to matters of body and soul. We can erect fences. We can say that side of the fence is body and this is soul, but it's all the same ground, really, the same arable earth.

An odd thing happened. I was reading an account of Lewis's relationship with Joy Davidman and there was something about loss being a "soul gap," as if the soul has been extracted like a tooth, and being aware of the gap, even, in fact, of *being* the gap. And this was said to be at the core of Lewis's experience of grief. As I was trying to make sense of the image I reached for an apple from the fruit bowl and, with the first bite, lost a tooth. So I sat back and played the tip of my tongue around the grief.

The Messenger Bird

ROB, WHO BOUGHT ME THE BEER THAT MADE ME MISS THE bus, and consequently my date with Melanie (I think that was her name), so setting a course for the rest of my life, married a warm, funny, beautiful woman called Josephine. She died of breast cancer in her forties. The morning after the funeral, as he and his children were eating breakfast, a bird flew in through the open window. It perched on the back of a chair, sang sweetly for a minute and flew off, leaving a deep tranquillity. Rob is a scientist with no belief in the supernatural, but he was convinced the bird was a messenger. Jo, his dead wife, was reassuring him: *Don't worry; everything will be fine.* And that's how it made them all feel. Everything will be fine. "I know it wasn't her," he said, "but I also knew that it was."

I hadn't seen Rob since he and Jo emigrated and went to New Zealand more than twenty years ago. We'd lost touch and I got to hear of Jo's death only through a colleague. Now here we were, Rob and me, sipping beer by the harbor in Auckland. Kate was back at the hotel recovering from the flight, but the cancer wasn't sleeping and I felt Rob's grief the more because I could glimpse my own, just up the road. There was a crack in his voice, I constrained a tear, and then we talked about sailing.

I know it wasn't her, but I also knew that it was. The words so casually violated a fundamental law of logic, the *law of contradiction.* Either the bird was a messenger from beyond the grave or it was not. It couldn't be both "true" and "not true" that the bird was a messenger. And if at some level Rob really thought it was then he would be relinquishing his rational, scientific beliefs for a spooky, supernatural view of the world that allowed for disembodied spirits communicating through birdsong. I knew that wasn't the case. He was still very much the scientist. The poet Keats called it "negative capability," being able to live with mystery, "without any irritable reaching after fact and reason."

You know, Rob said, it wasn't her, but, you know, I mean, you *know*. You just *know*. I know, I said. I know.

ON THAT FIRST visit to New Zealand my schedule confined me mostly to the city and, although we made a couple of excursions out of Auckland, I don't recall ever seeing the stars. Perhaps it was the weather. Three years later, out on the east coast of the North Island, I was taking a midnight stroll on a deserted beach where Kate and I had once strolled. The heavens were ablaze. Orion hung over the ocean, unreachably distant but intimate, as the constellations are. I know Orion like I know my face in the mirror, and saw something was wrong. I traced the core structures: the upper body defined by three stars at the belt, ascending left to right, with a bright star at each shoulder; the lower limbs marked by stars at each bended knee. But rather than hanging from his belt, Orion's sword was now held erect at chest height, as if in a ceremonial posture, and the lion's pelt, which should be draped over the outstretched left arm, had been discarded to lie at a distance from the hunter's right foot.

It took a fraction of a second to remind myself that I was in the southern hemisphere. I understood that the constellations have different orientations according to one's viewpoint on the surface of the Earth. Here in the South Pacific, diametrically opposite Europe, they would, to my eye, appear inverted. I was, so to speak, standing on my head relative to my customary view of the night sky. The framework of stars that gives shape to Orion's torso and lower limbs is fairly symmetrical, and that's what threw me. It looks roughly the same either way up. My brain reprogrammed itself and the heavens fell back into place, but the fleeting impression that Orion had raised his cosmic sword, that moment of perplexity, was exquisite. To witness the Great Hunter cartwheeling in the heavens gave the sense of seeing through alien eyes, of having traveled to a different part of the galaxy, not just the other side of the globe. The surest, most familiar things can suddenly become uncertain. It was an experience that left me with an acute awareness of the distinction between intellectual understanding and visceral knowledge. It's one thing to know that the constellations look different in the northern and southern hemispheres and quite another to witness the fact, *in the flesh.*

I took from my pocket a glass vial containing some of Kate's ashes and scattered them at the edge of the sea. Whatever else it is, grief is visceral knowledge. Grief is seeing the universe upturned, as if through alien eyes, the stars tumbling like kaleidoscope beads. Stabs of absence; stabs to the brain and heart; an entering of the flesh, a *knowing* in the flesh that she's not here anymore.

PART TWO

A Thousand Red Butterflies

Tabula Rasa

NINETEEN FORTY-EIGHT. IN A ROW HOUSE IN SOUTH PHILA-delphia curiosity gets the better of a small boy. He stands on the threshold of a forbidden territory. His father's voice is in his head.

"*Don't!*"

But he's already reaching, pulling the big black case, bigger than him, from under the parental bed. Little Pat opens the box and stares at the wondrous machine contained within, the metal and the wood, the graceful curves, the glistening strands of steel and wound bronze, the dark, hollow interior.

"*Don't touch . . .*"

His fingers are already exploring. They find stiff spikes of wire sprouting from the headstock. Soft skin is pierced.

"*Don't ever touch my guitar.*"

Pat walks calmly away, trailing blood graffiti on the walls.

There was something else in his head, other than his father's voice: a tumorous malformation of arteries and veins, a knot of worms about the size of an apple. It would be another thirty-two years before he knew about the rotten apple in his head and when eventually it was plucked from the left hemisphere of his brain the surgeon took with it a good two-thirds of the lobe in which it was lodged.

Pat Martino, the man the boy became, is sitting next to me. We're in his South Philadelphia home, his parents' old house as a matter of fact, and we're listening to the ethereal music of the great Estonian composer Arvo Pärt.

"Is this the one people die to?"

"*Tabula Rasa*? No. This is *Lamentate*."

"Oh."

I came to know Pärt's music through Fraser, a rock climber who liked to climb solo without a rope. "Deathbed music," he grinned, as we sat in his damp, cramped apartment, drinking tea and listening to *Tabula Rasa*. It was, he claimed, popular with the dying. I saw why. The simple patterns and progressions, the mysterious tones, and the perfect, unresolved silence at the end of the second movement seemed to lead close to the edge of being. Since falling from the great limestone wall at Malham Cove down among the Sunday strollers, Fraser had journeyed in and out of existence. His current, otherworldly self was not the original. His old employer didn't think so, and nor did his wife, but he was doing OK with a part-time job and a new girlfriend. The brain injury had dismantled his mind but he had reassembled himself in modified form and was happy enough with the new configuration. Music had been the blueprint and the glue. Through music he found connection with his old self, and reconciliation with the new. He gave me a CD, a random compilation of musical flotsam drifting from the wreckage of his brain: Bach; Coltrane; Dylan ("Knockin' On Heaven's Door"); and the deathbed music of Arvo Pärt. Of course it was anything but random. It charted the journey from Fraser 1 to Fraser 2.

I told Pat about Fraser. "Interesting," he said, and straightaway started downloading the music, three or four albums. I told him I had a theory, well, hardly a theory, more a vague notion, that one of the functions of music was to tune the brain machinery that drives our sense of self. There was music, like Pärt's, that worked to loosen the ego and there was music that had an opposite, consolidating, effect. Musicologists point to the collective functions of music, its use in ritual and ceremony and its contribution to the continuity and stability of cultures. Singing and dancing draw people together, synchronizing emotions, bonding the group in empathy and reflection, or in preparation for action. The power of music lies beyond language and intellect. It springs from an emotional need for communication with other human beings. But maybe there's something prior even to that. Music fills the body. It fuels the engines of emotion and action at the core of selfhood, the embodied self of the present moment. Without coherence at this level there is no possibility of developing a stable personal identity or social relationships. "Interesting," Pat said.

"You knew you were forbidden to touch the guitar?"

"Yes."

"Were you scared?"

"No. Once the decision was made I felt no fear."

Pat's father, Carmen "Mickey" Azzara, is long gone. We watch his image on the TV screen: the featherweight frame, the pencil mustache, the slicked-back hair. It's an old documentary and Mickey is retelling the fable of the guitar under the bed, voicing his son's words.

" 'Pop, I can play that guitar. I know I can play it.' "

Pat smiles.

" 'So, all right,' I said. I told him, 'you wanna play the guitar, I'll back you all the way. So go ahead, pick up the guitar. *Pick up the guitar!* . . . But don't you *dare* put it down.' "

Better the guitar than the trumpet, which could ruin a man's lips and make him unattractive to women.

I don't know if Pat really remembers the guitar under the bed. Sometimes he talks as if he does, sometimes as if he doesn't. But what does it mean to remember? It's not like reading a book or watching a documentary.

"Did I do something wrong, Mr. Paul?"

Mickey has taken Pat to the Steel Pier in Atlantic City to see a Les Paul gig. Les Paul, the Wizard of Waukesha, TV star, virtuoso musician and pioneer of the solid-body electric guitar. This would be 1956, I think. They've found a way backstage. "My son plays guitar, and he would like your autograph," says Mickey. "So here," says Mr. Paul to Pat, "take this guitar and play me something." Pat plays but the silence hangs heavy when he stops. Then Les says, "Who in the world ever taught you to play like that?"

"Is it wrong, Mr. Paul?"

Les is perplexed. He's never seen anybody pick a guitar the way Pat does, holding the plectrum as a polite young lady might hold a demitasse, pinkie extended. But the illusion of politesse had dissolved as soon as plectrum met string and started whipping up the notes with such astonishing fluidity and force. Les figures that sooner or later he is bound to hear of this lad again.

Three years later. Mickey is looking out the window.

"What the hell's that outside?"

"It's what we're going away with, Pop."

"With a *hearse*? They're taking you away in a HEARSE!"

It was a common mode of transportation for organ bands in those days. A hearse could accommodate a Hammond B-3 organ. There was no room in the front seat for Pat so he sat in the back with the B-3. He had with him a ukulele and played it all the way to Buffalo. This was his first road gig. He was fifteen years old. The bandleader, Charles Earland, was not that much older. I picture Charlie at the wheel of the ominous limousine, a smile and a wave for Pat's mom and dad. Charlie and Pat had been in the same high school dance band, along with a trumpet player called Francis Avallone who later morphed into Frankie Avalon, the teen pop idol. So much for Mickey's theory about trumpets.

The gig was at the Pine Grill. It was the place to be in Buffalo and, as Pat and the band were playing their set, in wandered Lloyd Price, the rhythm-and-blues singer. Lloyd was a star. Elvis had covered one of his songs, "Lawdy Miss Clawdy." During the intermission Lloyd called Pat over to the bar. "Listen," he said, "I like the way you play. If ever you want to come to New York City, just let me know." Pat called Mickey. "Pop, guess who asked me to join his band." "Take it," said Mickey. "Take it."

We switch off the T V. "That brought me to New York," Pat says. "We played forty minutes with Lloyd and before he came out we played an hour. It was an eighteen-piece band, the hottest big band in New York City at the time." A smile lights up his face. "And in the band were Charlie Persip, Stanley Turrentine, Slide Hampton, Red Holloway . . . So many great players, so many *amazing* players . . . and I was just a little kid—they called me *The Kid*—and I was just a little boy from Philadelphia who was in the middle of a dream."

IT'S COLD IN New York. It's April but there are snowflakes flurrying outside the hotel window. Caught in the upstream of the heating vents and other invisible eddies, they seem to be mostly falling upward. Out on the street it's freezing. I was here in December en route to Philadelphia to see Pat and had come prepared for a Manhattan winter, but it was warm as an English summer. Now, dressed for spring, I'm shivering.

I'm here with Ian Knox, a filmmaker friend. We're making a documentary about Pat. "You wanna reach out for Les?" Our eggs Benedict breakfast is interrupted by a call, and fantasy filters into the cozy, quotidian atmosphere of the Polish Tea Room on West 47th Street. Les? *Les Paul.* The Wizard of Waukesha. Yes, we would like to reach out for Les. We would, if possible, very much like to meet Mr. Paul. "You want *me* to reach out for Les?" We've been reaching out for Les for some time, in fact, but now the long arm of Joe Donofrio, Pat's manager, is offering to reach out from Atlantic City.

We make our way to the Colony Music Center at 49th and Broadway and ask permission to film. "No problem. Least I can do," says the man we take to be the supervisor. Meanwhile, two of his underlings are having a row. A door slams. "Do it your fuckin' self," one of them barks. The supervisor smiles serenely. When we return with the camera an hour later they're playing a track from

Pat's latest album, *Remember*. I try to make conversation with one of the sales assistants. They're mostly middle-aged men, fading as the sheet music and pop memorabilia fade around them. The sales assistant is tall and spare and there is something robotic about the timing of his movements. He hasn't understood a word I've said. It's my accent. "Yeah, whatever," he says. His metallic face switches off.

And then on again.

"Oh, you're the Pat Martino guys. Where you from? Australia?"

"No, England."

"Oh," he says. "How's Terence Stamp?"

"Don't know, haven't seen him for a while."

"I saw him the other day," the man says.

"Really?"

"In a Superman movie."

He switches off.

We cross Broadway and head a few blocks up to the Iridium Jazz Club. It's where Les Paul has agreed to meet us. He's into his nineties now but still does a regular Monday evening show at the Iridium. We can have an hour with him before the show. That's the plan, but there's a demon at the gate. The house manager is having none of it. Knows nothing about it. He's adamant. He's belligerent. We can't go through. He doesn't know who the fuck we are and he doesn't give a fuck. So that seems to be that. So there's me and Ian left on the stairs that lead down to the floor of the basement club, not knowing what to do. We're at the level of the mixing desk where a disheveled-looking guy, mid-sixties I'd guess, is sitting in the shadows eating a cheeseburger and fries. He's been listening. He introduces himself as "Rusty," and tells us to follow him. His dad would love to meet us. The house manager looks ready to burst a blood vessel but he's not standing in our way now. Rusty leads us down the corridor to the left of the bandstand, to a cramped dressing room where we are greeted by the diminutive, sprightly figure of his father, Les Paul, the Wizard of Waukesha. It's an easy conversation. Les is generous with his stories. He wants to get things right. "Pat's first wife," he says before we start, "you'll have to remind me. I can't remember her name. What was her name? I'm getting a little forgetful. Jenny? Josie? . . . Oh, yeah, Geri."

It turns out his memory is sharp. We hear of the night sometime in the early sixties when Pat introduced him to Wes Montgomery, who was playing an engagement at Count Basie's up in Harlem. Two other great guitarists showed up, George Benson and Grant Green, and the five of them, Les, Wes, Pat, Grant and George, ended up on the street corner at three in the morning and went on to Wells Chicken and Waffles for breakfast and talked guitar until the sun came up. Unprompted, Les takes us right back to the Steel Pier in Atlantic City, circa 1956, when young Pat was granted an audience backstage and invited to play, leaving the maestro dumbfounded by his virtuosity.

Les is generous with his memories, and he's generous with his time. The house manager, still furious, is now hovering in the background. Our time is up, but Les is having none of it. The stories keep rolling. He had no contact with Pat after their Steel Pier meeting but, a few years later, he hears of a prodigious young guitarist whose talent is scaring the shit out of other musicians, and he tracks him down to Small's Paradise in Harlem, and the redoubtable talent turns out to be Pat. The house manager is hammering on the dressing-room door now, but Les is unruffled. Finally, in his own time, he gets up to go. I notice he's forgotten his guitar.

So this is the Legend of Pat Martino. In the late sixties, with a catalog of successful albums to his name, and on the brink of a major new recording deal, Pat became seriously mentally ill. He was diagnosed with manic depression, which we now call bipolar disorder, but then the seizures started and it turned out he had an arteriovenous malformation in the left temporal lobe of his brain. It required surgery as a matter of urgency. The surgery saved his life but wiped his memory. He had no idea who he was and didn't even recognize his parents. The surgeon's knife had sliced selfhood from his brain. Nothing mattered; life was meaningless; he was nobody. So they tried to coax him back to being the *somebody* he once was. His father would play Pat's old records at full volume, which was an intolerable torment and, when he could stand it no longer, Pat fled. He drifted from New York City to Rome, to Barcelona, to the dark side of the Moon, through the heart of the Sun, to the outskirts of infinity, where, on some strange shore he strode through the storm, through the waves to the depths of the ocean, with Jimi at his side, not to perish, not to flee, but to be reborn from the womb of the sea, and Jimi said, it's OK man, have no fear, break free, you can breathe underwater, just follow me, we'll go find us some dolphins and hitch a ride to the ends of time, to the Isles of the Blessed. And if we end up at that other place you can just strum on a lyre, man, and melt the hearts of Hades and his lady, and they'll send you home to your mom and dad. Just don't look back! *Don't look back!*

NEWS CAME THROUGH—WHERE was he, Tokyo, Amsterdam?—that his mother had died, and he returned to Philadelphia. His father died soon after and it was all too much. Pat's life fell apart. Rage consumed him, and he ended up once again on a secure psychiatric ward. It was there that an astute psychiatrist gave him a primitive computer to play with. Its feeble 127KB memory contained a music program and Pat began to play, like a child with a toy. There was nothing to achieve, nowhere to go, forward or backward, nothing to do except play. It was an epiphany, like being born again, he said, but "living entirely in the moment." The music mattered; life was meaningful; he was somebody. And with

music as the golden thread, he began to weave a new version of himself. After years of silence, he took up the guitar again, studying technique, via tuition videos, from a great teacher: his former self. And, in due course, he ascended once more to the pinnacle of his art.

I wish the story about the tuition videos were true. It isn't, but even so, what a tale! A great virtuoso forgets how to play his instrument. A sublime talent destroyed. So what does he do? He starts over. He learns how to play again, from scratch, and he's right up there again with the greats. He's a genius twice over, but this time with a piece of his brain missing. Did it really happen that way? Yes and no.

If you want to know the difference between myth and legend, this is a good place to start. The Pat Martino story is a wonderful legend, but sometimes, the way it's told, it's more of a myth. So what's the difference? I think it comes down to the balance between meaning and facts. Let's say there's a continuum between, at one end, true, naturalistic stories based on factual events and, at the other, stories about supernatural beings and happenings. Legends fall in between. They are semitrue but may have supernatural elements.

If we look to ancient Greece, we know a lot of factual things about the people and the politics. We know about the organization of the city-states, we know about festivals and traditions. We know about military conflicts. We know, for example, that the Battle of Marathon took place in 490 BCE when the Persians invaded Greece, and that the Greeks were victorious. But the facts of the matter become blurred with legend. Herodotus tells us that the god Pan appeared on the battlefield, striking terror into the Persians and causing them to flee. Pan, a supernatural being, takes us right into the world of the Greek gods and goddesses, an entirely mythical world of supernatural, immortal beings whose lineage goes back to the origins of the universe and who preside over all aspects of human life. Over time, there was a gradual move away from literal belief in these supernatural entities, but they still gave meaning to people's lives. That's the function of myth, to give meaning. The facts, or lack of them, don't really matter anymore. It's all in the story, not the history.

Martino's story is a semitrue modern legend. There are factual components—the brain hemorrhage, the surgery, the memory loss, the abandonment of the guitar, the return, and so on—and there are legendary components, embellishments, misinterpretations, that give Pat's story a mystical quality. At the core of the Martino legend is the idea that he had to relearn the guitar totally from scratch. I didn't believe for a minute that Pat had literally forgotten how to play the guitar, not in the sense that he completely lost his memory for the instrument with a total loss of skills and knowledge. It just didn't fit with what we know about amnesia and about how the brain is organized.

There are different kinds of memory and there are quite different parts of the brain involved. So, for example, there's one broad division between "declarative" memory and "procedural" memory. Declarative memory is our long-term memory for facts and events. Procedural memory is more to do with skills. It's knowing how to ride a bike, how to type, how to play the guitar. Actors and dancers call it "muscle memory" but it isn't just about movement and dexterity. Procedural memory includes cognitive skills— being in command of the vocabulary of a language, for example. Reading music is another, as is the skilled musician's command of chords and scales.

You can lose your fact-based memory but keep the skills-based stuff, and that's the classic pattern in amnesia. There are cases of musicians with severe amnesia who forget they ever were musicians but who retain their musical skills even while denying they have any. The British musician Clive Wearing is a famous example.

In terms of brain anatomy, the skills involved in musical performance are laid down in older structures of the brain; that is, older in evolutionary terms—structures such as the basal ganglia and the cerebellum. But these areas weren't affected in Pat's case. Then there are other, "newer" areas involved in musical performance, such as the motor areas of the frontal cortex, but these weren't directly affected either.

So where did the idea come from, the idea that Pat had to relearn the guitar from scratch? He definitely seems to have experienced a period of amnesia post-surgery and it may be that he had no knowledge of having played guitar. By all accounts he showed little interest in the instrument; indeed, seems to have been quite alienated from it. His father's well-meaning but emotionally intrusive efforts to encourage him to play, and playing him his own recordings over and over, were counterproductive. Pat hated that.

At this distance in time it's hard to say exactly when the amnesia began to resolve and when Pat picked up the guitar again, or even if these phases coincided. I put these questions to his old friend John Mulhearn, also a guitarist. John spent time with Pat immediately before the surgery and in the weeks and months that followed. He couldn't say for sure what the tempo of recovery was, but I got the sense it was certainly weeks and possibly a number of months before Pat started playing again. And when he did, it was with a passion, playing and transcribing maniacally. John couldn't keep pace. So it was mostly a rediscovery of skills rather than a relearning, and a sudden resurgence of motivation. Even so, he had a long way to go to get back to his former levels of virtuosity, and his achievement in doing so is not to be underestimated. It was heroic in its way. Orpheus returning from the Underworld.

There remains the question of how Pat's psychological reactions relate to the brain damage he suffered. It's not at all clear why he should have experienced a loss of identity, because the key memory structures were not directly affected by the surgery. We did an MRI scan and it showed the hippocampus to be intact, left side and right. Severe amnesia, at least the sort associated with temporal lobe damage, typically requires damage to the hippocampus on both sides. In Pat's case it was only structures adjacent to the left hippocampus that might have been affected. I can only speculate. One possibility is that the surgery had indirect physiological effects on certain key brain regions, areas crucial for autobiographical memory, and that these effects subsided as the brain readjusted physiologically post-surgery. Maybe there were knock-on effects in the frontal lobe. Damage in one area can cause functional problems in other, healthy, areas. So maybe the removal of such a large chunk of Pat's temporal lobe indirectly affected frontal lobe areas that we know to be important for both motivation and autobiographical memory. But I'm hand-waving here.

As for the effects on Pat's memory over the long haul, it's difficult to disentangle genuine recall of remote memories from stuff he's learned since the surgery. He worked hard to relearn the names of family members and to reconstruct his life story through the accounts that other people gave him. I'm inclined to think that his autobiographical memories are a combination of the two, to some extent; stuff he never forgot and stuff he's learned since the operation. He appears to recall events from his early professional life quite fluently and sometimes in surprising detail. But there are some notable lacunae. Going through family photograph albums with him, as his parents had during his convalescence, I found there were occasions when the pictures drew a surprising blank, and his responses to some photos seemed oddly well rehearsed, as though he had learned the facts and faces by rote, rather than recalling information spontaneously.

His memory also lets him down in odd ways in real-life situations. To give one striking example, sometime around the mid-

1990s, when he had returned to performing, Pat played a show in New York City attended by the film actor Joe Pesci. The two had been close friends in the early 1960s but had lost touch and hadn't met in the intervening years. After the show, Pesci found his way to the dressing room and introduced himself, but it was apparent that Pat didn't remember him—or, rather, he recognized him as Joe Pesci, the famous movie actor, but not as an old friend. Yet the memories flooded back when Joe mentioned *Grasshopper*, a drink Pat used to enjoy. In Pat's words, "the moment he described the drink, a series of images appeared in my mind . . . I remembered the bartender and the stage and the position of the instruments that remained on the bandstand in between sets. And then I remembered Joe Pesci . . ." Pesci confirms the story.

Pat is also prone to occasional confabulation and the details of certain autobiographical anecdotes seem to vary, depending on the person with whom he is sharing them. For example, he tells a humorous story of once, on impulse, buying a Porsche sports car but, from talking to some of his old friends, it seems doubtful this ever happened. Others, though, are less skeptical, and one gets the sense that Pat has to some extent reconstructed his life history, his sense of who he is, through the not-always-reliable stories and memories of others, with consequent factual inconsistencies. His past, you could say, is indeterminate, like the past of those sub-atomic particles of the quantum world.

PAT HAS JUST played the Iridium. It's about two in the morning and a few of us are having a beer in the Rum House bar on West 47th Street. There's a piano player, an old acquaintance of Pat's, and she's getting people up from the floor to sing their party pieces, show tunes mostly. It's a good atmosphere. A strange-looking, sad-looking, pear-shaped guy takes the mic. He's dressed all in gray. He starts to sing. It's the most astounding voice, unearthly, a deep wobbly voice, getting deeper and wobblier as he goes. It could be an acoustic weapon designed to destabilize the rhythms of your internal organs. The bar falls quiet. People look at one another in disbelief. I look at Pat. He looks at me. It's a question of who's going to crack first. The piano player stops playing and she says to the singer, I think with genuine curiosity, "Are we doing the same song?" Silence, broken by a squeak, a honk, a snort, a prolonged wheeze. The sounds are coming from me. I am laughing like I haven't laughed in years. Helplessly. I am shuddering. Shaking. Weeping.

Egos and Bundles

I SHOW YOU A PHOTOGRAPH OF A FIVE-YEAR-OLD BOY IN school uniform and declare, "That's me!" when clearly it isn't me. I am a middle-aged man, not a little boy in his first year at school. It would sound contrived but I could instead say, "I was that boy," but then what does the "I" refer to? Descartes reached the conclusion that all he could be absolutely sure of was his own existence as a *thinking thing*, and he could even imagine this thinking thing as something distinct from his physical body. But if we play along with the idea that the "I" in "I was that boy" is an observing, remembering "thing," in short a thinking thing, it really doesn't get us very far. We're just kicking the can down the road because we have said nothing about the nature of the thing doing the thinking. An alternative view is that there is no "thing" doing the thinking. There is "thinking going on" (as brain activity) but nothing more to it.

Even if I cling to the idea of myself as a thinking thing, it's not at all clear in what sense I can consider myself to be the *same* thinking thing as the five-year-old. The boy and I are vastly different in terms of knowledge, skills, tastes and interests, and there is nothing more at the psychological level that I can point to as an intrinsic feature of our shared personhood. There are certain biographical and biological things we have in common. We were, for example, born to the same mother at precisely the same time—had the same birth, in short—and, not long after, were registered with the same name, and so on, but such facts don't link us at the level of Cartesian *thinking things*.

As for biology, we were built to the same genetic instructions, sharing a physical point of origin in the convergence of a particular sperm and a particular egg, but neither sperms nor eggs are thinking things. Genes don't introspect and there is no Cartesian "I" in the brainless bunch of cells that constitutes a fertilized egg. A Cartesian "I" had popped up by the age of five, for sure, one that was excited to get a shiny blue bicycle for its birthday, one that loved the scent of Miss Johnson, its first teacher. But I have no real idea as to when it ("I"; "He") came into being. The journey from egghood to personhood is a mysterious one.

My genetic match with the boy in the picture has determined us to have the same blood group and skin color, among other physical attributes, but we may doubt that these are unique, fundamental aspects of personhood. As regards the basic materials, the molecular building blocks of the body, the boy and I are not the same. Tissues are constantly regenerating. Old cells are discarded and new ones grow. With the passage of time there is a complete turnover of molecules. We are different lumps of matter. So what persists?

According to the philosopher Derek Parfit, there are, broadly speaking, two theories about what persons are and what is involved in their continued existence over time: Ego Theory and Bundle Theory. Ego Theory represents the intuitive, common-sense view that there is an "I," an experiencer of experiences that constitutes the essential core of every person. Descartes is the philosopher most closely aligned with this view. He believed our capacity for self-awareness is due to the possession of an immaterial soul (that *thinking thing*) and it is this that gives us coherence as individuals and continuity over time. Bundle Theory has origins in Buddhist teaching but owes its modern formulation to the eighteenth-century philosopher David Hume. It rejects the idea that actions and experiences are owned by an immaterial soul, or any other variety of essence, ego, or "I." There is no observing "I," just sequences of actions and experiences. Nothing more. According to this view, the self is no more than a bundle of fleeting impressions. Actions and experiences are interconnected but own-erless. A human life consists of a long succession of mental states rolling like tumbleweed down the days and years with no one (no thing) at the center. An embodied brain acts, thinks, has certain experiences, and that's all. There is no deeper fact about being a person.

Parfit devised a famous thought experiment that teases out the distinctions between Ego Theory and Bundle Theory. Imagine being teleported, *Star Trek*-fashion. Actually, not *Star Trek*-fashion, if Chief Medical Officer Dr. "Bones" McCoy's understanding of teleportation is to be believed. "Crazy way to travel!" he once declared. "Spreading a man's molecules all over the universe!" It doesn't really work that way. Here's how it works. A special scanner records the state of every cell in your brain and body and digitally encodes the information for radio transmission. Your body is destroyed in the process but reconstructed as soon as the radio signals are received and decoded at your destination. You "arrive" in precisely the same condition as you "left," identical in body, brain and patterns of mental activity, which is to say patterns of neuronal activity. Your memories, beliefs, plans, skills and emotions are perfectly intact and you go about your business feeling and believing that nothing about "you" has changed in the slightest. If you are comfortable with this scenario, then you should be comfortable with Bundle Theory. You appreciate that the observing "I" is no more than patterns of energy and information that can be disrupted and reconstituted without destroying the self, because there is no essential self to destroy. The patterns are all. If, on the other hand, you believe that some essential "you" would be lost in the process, then you are an irredeemable Ego Theorist. You believe that the reconstituted body is not "you" but a mere replica. The "replica" will believe deep in its bones that it is the very person who stepped into the scanner at the start of the journey, and friends and loved ones will agree. But, you insist, it could not be you because your body and brain would have been destroyed.

Incidentally, we see here a neat inversion of conventional thinking. Those who believe in some sort of essence, or immaterial soul, suddenly become materialists, dreading the loss of the original body. But those who don't hold such beliefs are prepared to countenance a life after bodily death.

A teleportation trip, Parfit suggests, is no more existentially threatening than "traveling" from one day to the next via dreamless sleep. In either case, the only thing that matters, in terms of what is preserved, is psychological continuity. You survive your night's sleep only because the bundle of mental states that automatically reconfigures on waking resembles the one that unraveled in the process of falling asleep. *Ah, but you wake with the same body*, some will say, as if that really matters. So here's an extension of the sleep scenario, which, at the same time, involves some radical (but possibly less threatening) "body replacement." You are offered a place on a spaceflight to Pluto. It takes about ten years to get there so you will be put in a state of suspended animation for the duration. The last thing you remember is stepping into the hibernation pod in Earth orbit before shooting off in the direction of the distant, dwarf planet. There follows a decade of dreamless sleep and the next thing you are aware of is being awakened as your spacecraft goes into orbit around Pluto, ready to descend to the surface. *Well, here I am*, you think. *It seems like only yesterday that we were preparing to launch from Cape Canaveral, and no time at all since I got into the hibernation pod!* But over the course of the journey, through natural processes of tissue regeneration and molecular replacement, there has been a steady turnover of the atoms that compose your body such that you are now made up of an entirely different collection of atoms than when you set off, as would have been the case if you'd been teleported. Can there be any doubt that you are still *you*?

A Thousand Red Butterflies

M ARCUS AURELIUS WROTE HIS *MEDITATIONS* WHILE SERV-
ing as emperor of Rome between 161 and 180 CE. They were
private reflections, never intended for publication, although they
now count as probably the most important source of ancient Stoic
thought. Marcus aspired to live each day as though it were his last.
You may leave life at any moment, he wrote. *Have this possibility in
your mind in all that you do or say or think.* His words came to me
one summer's evening on the island of Crete. There was cold beer
at my elbow and my sandaled feet were up against the trunk of a
pine.

This was the fourth year post-diagnosis. We didn't think we'd
get this far. The prognosis was poor. It was a malignant, invasive
carcinoma, Grade 3, we were told, and by the time it was detected
the disease had already advanced, in force, to the lymph nodes.
But here we were. Kate's hair had grown to a short crop in spiky
defiance of the last dose of the chemical poisoning that seemed to
be keeping her alive. The cancer had made inroads into the hip-
bones, which slowed her down a little, but the pain was tolerable.
She walked with a limp but swam gracefully.

We'd come through a tough time, more arduous for Kate, naturally, but something of a trial for me, too. So we splashed out
on a week at a five-star apartment hotel close to the sea. It was
called the Minos, in honor of the legendary king of Crete. I learned
about King Minos on my first day at grammar school, and about
Sir Arthur Evans's archaeological excavations at Knossos, which
had revealed the remnants of a once stupendous royal palace. Our
first homework assignment was to copy a textbook picture of a

figurine that Evans had unearthed, a snake goddess. I thought she was sexy. Each of her raised hands gripped a wriggling snake and there was a small, cat-like animal sitting on top of her head. She wore a long, flounced skirt with an apron, above which, and quite thrilling for an eleven-year-old boy, a tight bodice opened to present her bare breasts. I'd forgotten all about her, and now there she was staring back at me from my guidebook.

The book lay open in my hands but I wasn't reading. I was noticing colors: the pine bark running blue-gray to rust, the red geranium. I was noticing insects and animals: the tiny green bug on my forearm, the microscopic orange thing that dropped onto the book, no bigger than a full stop, the black cat stretching in the shade. The air was filled with the din of cicadas and Mediterranean scents. I sipped my beer and savored the moment. *One day I'll be dead.* It's an oddly exhilarating thought. Something unimaginable—eternal nothingness—awaits us all. The thought sharpened my senses, as the Stoics say it should. Glimpses of nothingness triggered a reflex grasp of the *somethingness* of sentience. We forget we're alive. If that geranium had suddenly burst into a thousand red butterflies it would be no more a celebration of consciousness than the dance of light and shade through the boughs of the tree. No less miraculous.

King Minos is said to have lived three generations before the Trojan War in the twelfth or thirteenth century BCE. According to one version of the story he was a son of Zeus, sovereign of the gods, and was raised by Asterion, king of Crete, whom he succeeded as ruler. The accession was disputed but Minos claimed divine support. To prove it he prayed to Poseidon, the god of the sea, beseeching him to raise a bull from the deep, and vowing in return to offer the animal as a sacrifice. Poseidon duly conjured up a fabulous white bull from the waves but, even though it secured his position as king, Minos reneged on his side of the deal, refusing to slaughter the magnificent beast. Poseidon was miffed by this and retaliated by infusing the king's wife, Pasiphaë, with an irresistible lust for the bull. She recruited Daedalus, the renowned master-craftsman, who constructed for her a hollow wooden cow on wheels, which, for greater authenticity, was covered in real cowhide. The contraption was wheeled to a meadow where the bull was at pasture and Pasiphaë climbed inside. The bull duly serviced her and the fruit of the union was the Minotaur, a monster with a bull's head and the body of a man. King Minos, by the way, was an inveterate philanderer. To put an end to his promiscuity Pasiphaë slipped him a potion that made him ejaculate snakes, scorpions and millipedes, which devoured his lovers from the inside.

WHAT ELSE WAS conscious in that summer evening scene? The tree? No. The bugs? I doubt it. The cat? I imagine so. I had an intuition that it felt like something to be the cat, that the animal had some awareness of the cacophony of the cicadas' mating calls; an awareness to which one might ascribe the sensory quality *sound*. As it stretched and rolled, I imagined it having a bodily sensation, which might be labeled *pleasure*, and I am pretty sure that if I had walked over and stamped on its tail then it would have felt *pain*. It was just an intuition, but perhaps it could be bolstered by biology. There is near universal agreement that consciousness is an aspect of brain function or, to be slightly more cautious, that brains are a necessary platform for consciousness even if, in themselves, they may not be entirely *sufficient*. The brain needs a body, and the body needs a world. I say *near*-universal agreement because some philosophers take the view that consciousness is a fundamental constituent of the universe, present to some degree as a feature of all material objects. This is known as *panpsychism*, and, if it were true, then the tree and the bugs would be imbued with the stuff of consciousness, as would the grains of beach sand between my toes. But I'll stick with the idea that consciousness is an aspect of brain function.

The cat's brain looks like a scaled-down version of yours and mine. That's not the case with the insects. A cicada's tiny brain is nothing like ours. I doubt they can "hear" the racket they make, even though it triggers impulses to act in certain ways. Consciousness, presumably, has evolved. But where on the evolutionary ladder does it first appear? Somewhere between the cicada and the cat, perhaps? But that's a guess. The seventeenth-century mathematician and philosopher René Descartes came to the conclusion that, lacking souls, *all* animals are unconscious automata. A more widely held assumption is that similarities between the nervous systems of humans and other animals make it reasonable to assume some degree of concordance in terms of basic sensory experiences, such as light and dark, heat and cold, pleasure and pain.

I can't be sure about the cat, the bugs or the tree but I do know for sure that *I* am conscious. It's a defining feature of my personal existence; *the* defining feature; the surest thing. *I think, therefore I am.* I know also that if a surgeon were to remove bits of my brain then my consciousness would be modified in certain predictable ways. According to the slice of the knife I might lose the capacity for vision, for hearing, or for abstract thought, and so on. The removal of still other parts of my brain would render me irretrievably unconscious. Such facts persuade me of a necessary link between my brain and the contents of my consciousness: the tree, the bugs, the thousand red butterflies. It's conceivable that I am the only conscious thing in the cosmos. To believe that would be a form of solipsism, which is the opposite of panpsychism. But given that all human brains are built to the same design and that other people do and say things consistent with sentience, I think it's likely they are just as conscious as I am. It would be madness to believe otherwise.

No one knows where consciousness first appears on the evolutionary timeline, and nor can we say for sure when it emerges in individual human development. I don't know the origins of my own consciousness. Like everyone else, I started out as a fertilized egg, cognitively more primitive than the orange microbeast traversing the page of my guidebook, let alone the cicadas. The egg wasn't conscious, I'm pretty sure of that, but I have no idea quite when the light switched on in my progression from embryo to fetus to newborn to infant. There is no agreed definition of consciousness, let alone procedures for determining its point of emergence in the developing brain. By four weeks post-conception, a part of the embryo called the neural tube has begun to develop three bulges corresponding to major structures in the mature brain: the hindbrain, midbrain and forebrain. Two weeks on and the cells are showing primitive electrical activity, less coherent, though, than that seen in the nervous system of a cicada. By twenty-three weeks the fetus possesses a complex and viable nervous system. Everything is in place for the emergence of a thinking, conscious human being, but there's no capacity for reflective self-awareness. The hardware is functional but the software has yet to be installed. Some psychologists and philosophers argue that conscious awareness reaches full bloom only with the acquisition of language and the consequent emergence of a reflective self, an observing "I." On this account, animals and pre-linguistic children lack self-awareness. Such people usually insist that neither is thereby morally demoted but, you wonder, does this make it more acceptable to eat small children or less acceptable to eat animals?

LISTEN, *the bull said to himself, nonverbally, I may be a beast of the field but I'm no mug. I'm doing this of my own conscious volition, in full cognizance of the fact that the object they just wheeled into my field, the object with which I am about to couple, is not a real live cow. The woman who climbed inside, all aquiver (do they think I'm blind!) need not have bothered. I'd have entertained her anyway, without the ridiculous subterfuge. That said, it's a piquant combination: ersatz bovine with a beautifully crafted, leather-lined slot and the succulent mysteries of the king's wife within! Of course, lacking language, I could not communicate any of this to the humans. So, Hup! Huff! Gnufff! Mahoooo! . . . Ah, comme l'affichage d'une lettre!*

The Consciousness Club

W E'RE AT THE WORLD'S END, DOWN BY THE FERRY PORT. Lewys is on the brink of tears and I don't know what to say. Ava will join us once she's set the babysitter up. It's Ava he's upset about. Did he have any suspicions? No, he says, why would he? It was a stupid question. Indeed, why would he? That's the thing. Is he sure? Yes, he's absolutely sure. So what's he going to do? He doesn't know what he's going to do. He's confused. That's why he's here talking to me. And what does Ava say? She says it's a load of rubbish. It's insane. He's losing his marbles. Lewys blows a sigh. *Fuck*, he says. *What the fuck am I going to do?*

Ava arrives, and acts as if everything is normal. She's Dutch and we greet each other Dutch-style. Three pecks, left, right, left again. I tell her she's looking well, which she is. She ruffles Lewys's hair and pouts at him mockingly. Then she turns to me and rests a hand on my arm. Her dark Dutch eyes look deep into mine. *Paul*, she says. *Me! Who'd have thought?* She goes to the bar and buys a round of drinks. We watch her joking with the barman. You still love her, don't you? I say to Lewys. He shrugs.

I GOT TO know Lewys, a swarthy, foul-mouthed Welshman, through the Consciousness Club, where, in a back room of The World's End, we downed gallons of ale and got into vociferous wrangles over artificial intelligence and weird neurology. Lewys is a computational neuroscientist. I'd been thinking about con- sciousness forever, but for Lewys it was a newfound passion. Till then, he'd been coasting along doing rather dull research on computational models of sentence construction, dull to my mind, anyway. But what do I know? He was publishing in high-impact journals and hauling in more grant money than the rest of us put together. In the eyes of the university, Lewys was a supernova. But then he discovered Faraldo's Theory of Neuronal Relativity, since when he had become obsessed with the problem of consciousness and hadn't published a thing. This was four or five years ago. Fair enough, it takes a while to get research up and running, and papers into print—let's say a couple of years, if all goes well—but, as time went by, it became clear that Lewys had no intention of publishing anything. He slipped into a low-grade depression—flat, fatigued, taciturn, irritable—and it was just before Christmas last year that I found him at his desk, first thing in the morning, staring catatonically at his computer screen. He'd been there all night. I couldn't get anything out of him. When, finally, he did speak, the voice was a funereal drumbeat. *Faraldo, Faraldo*, he intoned. *That theory of yours is Pandora's fucking box. I've opened the lid and the demons are flying. Look!* he said, pointing to the screen. *There's one of the fuckers.* All I could see was the image of a brain, roiling like molten copper.

The late Paolo Faraldo, often referred to as the Einstein of neuroscience, was a polymath: genius logician, Fields Medal–winning mathematician, and world-class neuroscientist. As if these weren't strings enough to his bow, he was also an accomplished poet. It was his poetic sensibility that first drew him to the mystery of consciousness and, ultimately, to the unraveling of that mystery. He came to realize that lyric poetry is the closest we can get to the description of *qualia*, the ineffable, irreducible raw essences of conscious experience: the redness of a poppy, the warmth of the sun, the ache of unrequited love. Other art forms manipulate qualia. Music works with sequences and structures of sound. Paintings play with the effects of line, light and color. But poetry is a portal through which we glimpse the shimmering inner core of sentience itself. As Einstein imagined himself riding a lightbeam, Faraldo pursued trains of poetic thought through networks of neurons, and arrived, in time, at a true view of consciousness. Traveling at the speed of light, young Albert saw that space and time are fused in a single continuum. From the depths of the poetic brain, Faraldo saw a similar fusion of mind and matter.

Faraldo's solution to the problem of consciousness was so elegant, so mathematically precise, that those who grasped it came to wonder why they ever thought there was a problem in the first place. There were skeptics and dissenters, of course, who sneered that Faraldo had not so much explained consciousness as "explained it away," or even that he was denying its very existence. Far from it. Human beings were obviously capable of finely tuned awareness of themselves and the world around them. The great apes, elephants, dolphins, dogs and birds all possessed varying degrees of sentience and even, in some cases, self-awareness. Other living things did not, for example mushrooms and microbes, and there had long been debate as to where on the evolutionary scale the consciousness cutoff might be. Faraldo exposed the pointlessness and futility of such debate.

Lewys had presented Faraldo's work at a meeting of the Consciousness Club, and I'd joked that my wife had sorted it all out well before Professor Faraldo. *What's the problem?* she'd say. *Honestly, I don't get it*, and I'd find myself hard-pressed to say what the problem was. I would flounder. Well, I'd say, it's the problem of explaining how and why physical states of the brain produce mental experiences. Just how is it *possible* for mental stuff—the blueness of a blue sky, the bitterness of lemon juice—how is it possible for that to arise from physical stuff? *What do you mean*, she'd say, *"arise from?"* Well, what *did* I mean? She said I was creating a problem where there really wasn't one. The blue of the sky and the bitterness of lemon juice was just the brain getting on with doing blue stuff and lemony stuff. OK, then, *Why? Why* consciousness? By now she would be losing patience with me. Kate had no time for metaphysics. No, wait, listen, *listen*, we can figure out how the brain works. Right? We can map out different systems and, the thing is, we know they mostly work automatically and unconsciously. So, the question is, why the added extra of consciousness? Hmm? I have to think hard when I play chess. It takes a lot of conscious deliberation, but the chess program on my computer beats me effortlessly. *Unconsciously.* So, if consciousness isn't required for playing chess—or auto-piloting a plane, or solving equations—well, why should it be involved in anything else? *It's too late to cook*, she'd say. *Let's get a takeaway.*

Few had the breadth and depth of knowledge to grasp Faraldo's solution in its entirety, perhaps a dozen scholars around the globe. Lewys was one of them. For the rest of us it was a case of, well, you'll just have to take our word for it. The problem of consciousness, my friend, is no longer a problem. I suppose it's a bit like taking the word of a quantum physicist when he tells you a subatomic particle can be in two places at once. The idea seems absurd but those chaps have done the maths and they've done the experiments, and that's just the way the world is, whether or not we can get our heads round it. Lewys, though, understood the full implications of Faraldo's work and it consumed him. Faraldo rather dramatically consumed himself, committing suicide on the day of publication of his seminal paper, *Neuronal Relativity and the Problem of Consciousness, Part 1: The General Theory.* Part 2, which would have provided experimental support for the theory, never appeared. The great man had apparently fallen into a sudden and profound depression. He left his lab one blustery November night, made his way to Russell Square tube station, and threw himself under a westbound train. Before doing so, he had fastidiously destroyed six months' worth of closely guarded notes and experimental data. He could obviously see what was coming, Lewys said.

What really hooked Lewys was a corollary of Faraldo's theory that he and a small handful of his peers saw straightaway. Sentience is *diagnosable.* It should, in principle, be possible to devise a practical method of detecting consciousness. This had important medical implications. It would, for example, be possible to know for sure whether someone in a seemingly permanent vegetative state had any degree of conscious mental life. In the past, life support systems had been switched off in the mistaken belief that the patient was deeply and irretrievably unconscious when, in fact, they were aware and concerned by everything going on around them. It would also settle the issue of assessing levels of awareness under general anesthesia, a common, if less commonly discussed, problem for surgeons and anesthetists. (The prospect of open-heart surgery is worrying enough without the thought that you might return to consciousness just as the surgeon is burrowing into your chest. It happens.)

Lewys discovered that consciousness came down to the presence or absence of certain highly specific patterns of brain activity, in particular, interactions between the sensorimotor systems, including the cerebellum, a subcortical relay station known as the thalamus, and activation in a deeply enfolded area of cortex called the insula. The signature of consciousness was complex, but it was consistent and recognizable using fairly standard functional brain imaging methods. Lewys had cracked the code. He had, in effect, invented a consciousness meter capable of detecting the merest whispers of sentience. This was scientific achievement of the first rank but, still, he wouldn't publish. There was something bugging him. He didn't say what.

This is what. I got a call from Lewys late one night and he poured it all out. I was already in bed and asleep when the phone rang. I'd drifted off listening to Proust on audiobook; it's a potent sleeping draft I often take. I now know the "Overture" almost by heart but never get much beyond that, at least not in a state of wakefulness. It was the middle of the night and, like Marcel, I found myself not knowing quite where I was. Half in Combray, it seemed, because the Frenchman's thoughts were still laboriously unfolding in the background. I switched him off. Now it was

Lewys's rambling thoughts I had to contend with, and, as I listened, I began to fear for his sanity.

"Lewys, are you all right? You sound very pissed."

"Yes, mate, I am, but what I'm telling you is the truth."

"Have you had anything else?"

"What?"

"Weed?"

"No."

"*Anything?*"

"NO! Listen to me! Just fucking listen!"

He was angry now and it seemed to sober him up a little. He told me the whole story again. Here it is. In the process of developing his consciousness meter Lewys had made a truly startling discovery: not all human beings are sentient. Some fully functional people operate in an entirely nonconscious state. At first he thought it was a statistical artifact. But it wasn't. He worked days and nights on end trying to track down possible hardware and software glitches. There were none to be found. His results were indisputably valid and reliable. *Not all human beings are sentient.* Roughly speaking, around 10 percent are so-called philosophical zombies. We are not talking here about the shambling undead of those 1950s B-movies and horror comics. Philosophical zombies are an altogether more subtle creation. A philosophical zombie is conceived and born in the usual way, and it grows up and does the same things as the rest of us. Its brain looks like yours or mine and operates in pretty much the same way. The only difference is that the zombie is insentient. Throughout its entire life it has not the slightest flicker of conscious experience. Zombies laugh and cry just like you and me. They yell and curse when they stub their toe. They say they love you, they say you piss them off. They make orgasm sounds when they're having an orgasm, and sometimes when they're not. There are straight zombies and gay zombies, liberals and conservatives. There are Muslim zombies, Christian, Jew, agnostic, atheist, zombies and every other flavor of belief. Zombies get married and they get divorced. They get lyrical. They say they adore sunsets and bright autumn mornings. They read poetry. They listen to Mozart, eyes closed, beatific expression on their face. They do all this but they don't feel a thing. They say they do but they don't. It's just words. The lights are on but there's nobody home.

Philosophical zombies had, until now, been entirely hypothetical beings, invented by philosophers as characters in thought experiments whose purpose was to test the logical limits of materialist conceptions of mind. I'll spare you the convoluted details of the debate but, if zombies are at all conceivable, the argument goes, then there must be more to the mind than the physical activity of the brain. Now Lewys was telling me that zombies were real, not merely conceivable. They walk down every street. Ten-to-one your best friend is a zombie. He kept his startling discovery to himself. He was inclined to destroy the data. This was dangerous knowledge. The world was not ready for insentient humans. This was *Blade Runner* turned inside-out—insentient humans among us as opposed to sentient androids.

In the following days, Lewys and I spent long, coffee-drenched hours in his lab, poring over the data, turning his results this way and that, in search of an alternative, less gothic, interpretation. This was for my benefit because Lewys had been thorough in his work and already knew, for sure, that there was no plausible alternative. He took me through it all, step by step, and I could find no fault with the design and execution of his experiments, or the logic of his arguments, or, despite my reluctance to accept it, his conclusion. Some people are insentient. It's a plain fact, just as some people are left-handed or homosexual or blue-eyed. Insentience is part of the rich pattern of human variation.

And then he developed a terrible, nagging doubt. You'd think it would have occurred to him straightaway, but no, it was a while before the idea took root. Maybe the gatekeeper of his own conscious awareness wouldn't allow it through at first, but it came to him one night when he and Ava were making love. He looked into her eyes and had this thought: might my darling Ava be one of the insentient 10 percent?

AVA RETURNS WITH the drinks. "I thought he was joking," she says. "Well, he *was* joking. He was making a joke of it, even though he was deadly serious, weren't you, my love? I think I should get you into the scanner, you said. Let's make sure *you're* not a zombie. Ha ha ha. I told you not to be so ridiculous, but you wouldn't let it go. You went quiet on me, miserable, like your Welsh weather. Anyway, in the end, I thought, why not?"

Ava tells me she went through some cancer screening not so long ago. Lewys was with her when she got her results, and the news turned out to be good. She mimics Lewys's sigh of relief, his closed eyes rolling up to some inner heaven behind the eyelids. Not this time. This time, when the results of the consciousness screening came through, Lewys had a blank stare and the blood drained from his face.

I'm not much in the mood for drinking, nor is Ava, who has taken just one small sip from her half pint. Lewys, though, is now on his way back from the bar with a third beer in his hand. Have they told anyone else? No they haven't. Why me? They trust me. It's not just that I'm a friend. Lewys values my clinical expertise. *You've spent your entire working life dealing with . . .* (he struggles for words here) . . . *atypical brain function.* Ava rolls her eyes. She sighs. *Lewys,* she says, *we are going to carry on as normal. Nothing has changed. Really. Paul will explain this to you, won't you, Paul? Paul will tell you all this is a bunch of nonsense. It's a bunch of nonsense because, believe me, whatever your brain scanner is telling you, I am conscious, and all this, Lewys, is starting to upset me. It's not my brain you should be worrying about. It's yours. That's right, Paul, isn't it?* I don't say anything, but have a thought that instantly feels unworthy: *well, she would say that, wouldn't she?*

Any neurologist would look at Ava and tell you she is perfectly normal. There are no symptoms. Routine clinical assessments would find no signs of neurological disorder. Standard brain-imaging procedures would show nothing out of the ordinary. Ava is normal. Ava is an intelligent, life-loving, thirty-seven-year-old woman, devoted to her husband and their adorable two-year-old daughter, Agatha, whose second name (Lewys's bad idea) is Qualia. I've seen Ava laugh and I've seen her cry. I once saw her fly into a fury over something someone said. How could this clever, passionate creature not be sentient? Lewys is a brilliant scientist. I can't find fault with his work. And yet, and yet, and yet. Zombies? Seriously? To quote the old miserablist Schopenhauer, *The heart rebels against this, and feels that it cannot be true.* Or is it really the way things are, and, like the denial of death, another case of feeling something to be false (death is inconceivable/zombies are impossible) and knowing it to be true (death is inevitable/actually, yes, zombies really do exist)?

Yet, despite the quality of his science and the clarity of his arguments, I can't help feeling Lewys's present state of mind has the ring of delusion about it, a touch of the Capgras. In Capgras syndrome, the person develops a delusional belief that someone close to them has been replaced by a double. One explanation is that the brain's face-recognition and emotional-response systems have become disconnected. When we see someone very close to us, we experience an emotional charge at some level. In the case of Capgras patients, the theory goes, face recognition happens as normal but doesn't trigger the usual emotional reaction. The brain then leaps automatically to the false conclusion that the person you are looking at can't be the person it knows and loves. Another suggestion is that, ordinarily, the brain holds two quite distinct representations of a person. One registers their physical presence: face, body movements and mannerisms, voice, and so on. The other representation is more psychological. It frames an image of what the person is like on the "inside": their beliefs and preferences, ways of thinking and emotional characteristics—their personality, in short. In Capgras syndrome, the internal representation is blocked. From the outside, the impostor looks identical to the loved one, but sufferers are convinced it's not them on the inside. Incidentally, I've always thought Capgras syndrome was a perfect metaphor for falling out of love. The person looks the same, but a certain inner something is irretrievably missing.

I suppose the position Lewys and Ava find themselves in is Capgras-esque. In Lewys's eyes, Ava is not the same on the inside. In fact, she is absolutely nothing on the inside. The difference is, he doesn't think there's been a switch. He believes Ava must always have been that way. It's Lewys's understanding of her that has altered, and it is not a delusion. His new perception is a true perception. Ava is verifiably insentient. The thing is, it doesn't feel that way to me. When I am in the company of my friend Ava, when I engage with her words, and smiles and sparkling eyes, it feels to me like I'm in the presence of a warm, vibrant, fully sentient human being, even though there is no conscious mental life behind the words and smiles and sparkles. But, I ask myself, is this really any different to what we all do all the time in seeing a soul or a self behind another's eyes when, the truth is, there is no such thing, just a brain doing its silent, soggy, material stuff. So, no, I didn't see Ava any differently.

"Lewys, this is the maddest thing I've ever heard."

"But it happens to be true."

I didn't see Ava any differently, simply *couldn't*, but Lewys did and it was too much for him.

I'M SITTING ON a bench down at Devil's Point, where the river meets the sea, the place I used to sit with Kate. We were here the week before she died, a thousand years ago, yesterday. The tide is turning, and the Moon, far away through the blue, blue sky, is pulling the sea upriver. A thought crosses my mind. Maybe Kate was one of the insentient 10 percent? No! It's a preposterous idea. I'm laughing out loud and the passersby think I'm mad. It's time I made my way.

Last night I got a surprise call from Lewys, his first contact in two years. He said he had some wonderful news, but wouldn't say what. He just said let's meet for lunch at The World's End. So, it's on to The World's End, where I find Lewys and Ava sitting out in the sunshine, together again. I've never seen them looking so happy. Over fish and chips, and hoppy beer, Lewys tells me the story. He had traveled aimlessly for a while, India, South America, the usual, before accepting a post at UCL in Faraldo's old department. Ava had gone back to live with her parents in Nijmegen, before moving on to Rotterdam. Well, that much I already knew. Lewys had given up on consciousness studies. He was back to his old, dull research into syntax. But not for long. A former colleague of Faraldo mentioned casually over coffee one morning that he had stumbled upon a draft of Part 2 of Faraldo's seminal *Neuronal Relativity* paper, long thought to have been destroyed, and he wondered if Lewys would care to take a look. It was an irresistible offer.

The main body of the paper had, as Lewys had always suspected, anticipated his own work on the development of practical methods for the diagnosis of consciousness. Faraldo had, also predictably, discovered the reality of philosophical zombiehood, arriving at a remarkably similar estimate of the percentage of zombies among the general population. But the paper ended abruptly, with no general conclusion. Instead, appended to the unfinished manuscript there was a batch of notes, handwritten, apparently in Greek. Lewys showed them to a Greek colleague but she couldn't make head nor tail of them.

The language turned out to be Tsakonian, a descendant of the language of the Spartans, which is still spoken by a few hundred, mostly elderly, people in the eastern Peloponnese. Faraldo had no Greek family as far as anyone knew, and there's no call for Teach

Yourself Tsakonian packages, so presumably he had a teacher. Try Tavoularis, someone suggested, and Thanos Tavoularis, emeritus professor of classical languages and literature at the University of Oxford, turned out to be the man. Lewys paid him a visit at his ramshackle cottage in Boars Hill. Tavoularis was a week shy of his ninetieth birthday, but still as sharp as a Spartan spear. He already knew the general thrust of Faraldo's work, so had no need for a tutorial, and he was unfazed by the notion of zombiehood. He'd been a friend of the classicist Bruno Snell, who had argued persuasively that interior mental life was a product of cultural evolution. According to Snell, it simply did not exist, as we know it, prior to the intellectual revolutions of classical Greece. From his conversations with Snell, Tavoularis had come to appreciate that consciousness should not be taken for granted. Indeed, it was a common interest in consciousness that fired his friendship with Paolo Faraldo. *And this*, he said, holding up Faraldo's Tsakonian notes in an almost celebratory gesture. *Well! Well, well, well . . .*

Faraldo, like Lewys, had deciphered the neural signature of consciousness and had duly developed procedures for its verification, subsequently discovering, as Lewys had, that not all humans are sentient. But then he went a step further. He put his own brain to the test and the result was existentially shattering. He discovered that he was himself one of the insentient 10 percent. Faraldo was a zombie. It was all there in the Tsakonian notes.

"Delicious," Lewys says, pushing his plate away.

"That's all very interesting, Lewys, incredible, actually. You've got to write this up. But you still haven't told me how you and Ava got back together."

They look at one another. They look at me. "Go on," Ava says, "tell him the good news." Lewys exhales, cocks his head, presses a wistful smile, and says, "I'm one as well. I'm a zombie, and, you know what, mate, I couldn't be happier."

I tell him he's taking the piss. They both are.

"No," he says, "it's true. I'd never bothered before because, well, why would I? Why would I put myself through the sentience diagnostics? I mean, fucking hell, why would I? Of course I'm conscious! I'm telling you I am, I'm telling myself I am. But once I'd taken on board what Faraldo had done, the deep meaning of it, the fucking irony, I thought, why not? In fact, I *must*. So I did. And it turns out that I am, after all, not conscious, despite telling you I am, despite telling myself I am, and despite truly believing I am. It broke Faraldo, but for me it was a liberation. Faraldo's first thought was suicide. My first thought was Ava, Ava my soulless soulmate. I realized how much I was missing her, but what was there to stand between us now? We are kindred non-spirits! And I'm telling you, it feels wonderful, even though it's as sure as shit, as sure as the Earth revolves around the Sun, that I don't feel a thing! I am, though, telling you it's wonderful. That's the thing. *That's the thing!*"

Ava says, "Remember when that rock star and his film star wife split up, and they said it was a 'conscious uncoupling'?"

"Yeah, well, fuck that!" Lewys says. "We're getting back together and it's a fucking unconscious coupling."

He leans over and kisses me on the cheek. Ava kisses the other one. Tears are rolling, theirs and mine.

Rotten to the Core

"**W**HAT MAKES YOU THINK YOU'RE DEAD?"

Martin, a pallid, paunchy, fifty-five-year-old florist, has his elbows on the table and his head cupped between his palms. He is looking out of the window. It's taking a while for him to reply to my question so I carry on scribbling notes. *Notably slower today. Drowsier, dreamier.*

"I . . ."

Now I'm doodling a little picture of Martin. It looks a bit like *The Scream*, drained of angst.

"Because . . ."

He's thinking it through, and his response, when it arrives, sounds clear and considered, like he's giving me the plain truth.

"It's because I'm nothing now. I don't exist anymore."

Our deepest, our most fundamental, intuition is that we exist. For Schopenhauer, existence was an astonishing fact. For Descartes it was the only thing in the world that, at logical rock bottom, we could know with any certainty. We can't trust our perceptions. The whole world, even our own bodies, could be an illusion conjured up by an evil demon. But we can be sure of our existence, he argued, by virtue of the very fact that we are capable of thinking about whether or not we exist. *Cogito ergo sum.* I think, therefore I am. We may not be able to say with any certainty or precision quite *what* we are, but we know, at least, *that* we are.

Now here was a man declaring his nonexistence. Martin had a variety of symptoms, including drowsiness, problems with balance, forgetfulness and poor concentration. My tests clarified the cognitive problems: poor retention of visual information, dysfluent and inflexible thinking, and impaired spatial awareness. Applying crude rules of thumb, it was a pattern that possibly implicated the frontal lobes of the brain and the posterior right hemisphere, but it wasn't diagnostic of anything in particular. (He was, in due course, diagnosed with a rare autoimmune disease.)

Cognitive difficulties noted, Martin was otherwise well oriented, engaged and communicative, and able to give a clear account of his life story. He told me about the schools he'd attended, could reel off the names of teachers and classmates, and was similarly forthcoming about his employment history and family life. I wouldn't say he was animated, but his face lit up when he told me he'd recently become a grandfather. And yet, here he was telling me, matter-of-factly, that he thought he had died. Perhaps he was speaking metaphorically, I wondered. This was his way of expressing a feeling that because of his illness he was no longer his "old self." But, no, he meant what he said and believed himself, literally, to be dead. This is a rare and strange delusion, known as Cotard's syndrome.

In 1882 the French psychiatrist Jules Cotard published a series of case studies of people suffering what he termed *le délire de négation*. His patients varied widely in the details of their clinical presentation but all had self-negating delusions of some sort, ranging from beliefs that parts of the body were missing, or rotting, to a complete disavowal of bodily existence. The death delusion usually occurs in the context of severe depression. I first encountered it as a newly qualified psychologist working in a psychiatric unit. The patient was a profoundly depressed woman who told me she had been dead for some time and that I might as well have her buried. She believed her insides had already rotted away. *I am rotten to the core.* But Martin was not depressed, and nor was there any history of mood disorder.

Adam Zeman, professor of cognitive and behavioral neurology at the University of Exeter, invited me to see another Cotard's patient, one who subsequently was to be the subject of the most significant neuroimaging study of Cotard's syndrome to date. Graham was forty-eight years old and had no medical history of note other than a brief depressive illness. He was referred to a psychiatrist after a suicide attempt by electrocution. Eight months later he told his general practitioner that his brain had died and that, "I am coming to prove that I am dead." Despite my efforts to persuade him otherwise, presenting the plain evidence that he was fully engaged in a conversation with me, Graham was unwavering in his conviction that his brain had died the day he plunged a live electric heater into his bathwater. I queried how this could be possible because, surely, he would accept that a living brain was necessary in order to be able to think and speak. It was a bit baffling, he conceded, and he didn't have a good explanation for this strange state of affairs, but the fact was his brain was dead, and there you go.

Graham was referred to the University of Liège for specialized brain imaging and there, under the supervision of Professor Steven Laureys, he became the first Cotard's patient ever to undergo a positron emission tomography (PET) scan. PET scans provide detailed, 3D images that can indicate regions of abnormal brain function. The results were remarkable. Graham's overall gray matter metabolism was 22 percent below normal. Gray matter is the darker tissue of the brain, made up mainly of nerve cell bodies and their branching dendrites, whose job is to receive impulses from other cells. It's where the thinking goes on. There was also an intriguing difference between cortical and subcortical areas, that is, between the upper regions of the brain and various lower structures. Networks at the top end, in the cerebral cortex, were abnormally quiet, whereas the subcortical zones were buzzing. It suggested, according to the published report, "a profound disturbance in brain regions responsible for 'core consciousness' and our abiding sense of self." Steven Laureys later remarked that he'd been analysing PET scans for fifteen years and hadn't seen anything like it. Graham's brain function resembled that of someone under anesthesia, but there he was, wide awake, on his feet and interacting with people. The disjunction of brain pattern and behavior was unique.

WHAT IS IT like to experience the Cotard's delusion? Some of Martin's experiences seemed to resemble the relatively common psychiatric symptoms of "depersonalization" and "derealization," which are usually associated with mood and anxiety disorders. Depersonalization is a disturbance of self-awareness such that the person feels detached and unreal, often with a muting of sensory experience and emotional numbing. Derealization denotes similar experiential distortions of the external environment such that other people seem lifeless, objects lose their significance, and the world generally seems "less real." Given that both conditions involve sensory changes and feelings of unreality, they might be considered to be two sides of the same coin, and in fact depersonalization and derealization typically do co-occur. They are, however, dissociable. Depersonalization sometimes presents without derealization, and vice versa.

Martin could be said to show features of a depersonalization-derealization syndrome, but lack of insight into his state of mind excluded the diagnosis. People with mood- and anxiety-related depersonalization often speak in terms of feeling as if they are unreal, or that the world seems remote and unreal, as if being viewed through a glass wall, but the "as if" is crucial. They retain insight into the fact that their sense of unreality is illusory. Martin, on the other hand, was convinced that he and the world he inhabited were lifeless and insignificant to the point of true oblivion. The fact that we were sitting together having a conversation did not persuade him that he was alive and that the objects and events of the world around him were real. Nor did his *thinking* that everything was unreal constitute evidence of his existence. In this regard he clearly departed from Descartes, whose *I think, therefore I am* was proof that, whatever else could be doubted or denied, his own existence as a thinking thing was indubitable. I pressed Martin on this a few times and his responses were along the lines of "my thoughts aren't real either."

Both Martin and Graham were clearly deluded, but what was the nature of their conviction that they were dead? Delusions have been defined straightforwardly as false beliefs held in the teeth of contrary evidence, but there is a growing body of clinical opinion that delusions should be considered not as beliefs but as knowledge claims—distinguishing "to know that" and "to believe that." The deluded person simply "knows" such and such, rather than merely believing it. For example, I *know* that my name is Paul. It's on my birth certificate, it's what my parents always called me, what my friends know me by, and the name I give when strangers ask. On the other hand, I merely *believe* that my new neighbor is called David. That's what another neighbor told me, at least I think that's what she said. If, when we eventually meet, he tells me, no, his name is Dennis, not David, then I would accept this new evidence and henceforth call him Dennis. Nothing could persuade me that my name is not Paul. Delusional statements, as knowledge claims, are expressed with absolute conviction and certainty. They are not susceptible to rational interrogation and evidence-based counter-arguments. They are rock solid in the way that knowledge of one's own name is rock solid. Martin and Graham simply *knew* they were dead, as surely as they knew their own names.

COTARD'S SYNDROME IS a disturbance of self-awareness whereby the normal intuitions of embodiment and in-the-moment consciousness seem to be severely undermined. Other disorders affect selfhood in other ways. Memory loss, for example, can undermine the continuity of identity. Now, we know that the mind is anatomically distributed. The various components of mental life—perception, language, memory, and so on—are delivered by distinct brain systems. To descend into jargon, the mind can be "fractionated" neuropsychologically. In other words, different disturbances of brain function have specific effects on mental function. Memory, for example, has dozens of neuropsychologically distinct components: encoding, storage and retrieval processes; short-term, long-term; verbal, visual-spatial; recognition, recall; memory for facts; memory for faces; skills memory; autobiographical memory; and so on. "Mind" and "self" are interlocking concepts, so it seems reasonable to ask whether the components of selfhood are dissociable in the same way, or would be if we had some sound theoretical guidelines for thinking about the "components of selfhood," as we do for the components of memory.

The philosophical Ego and Bundle theories we discussed earlier (page 118) don't get us very far in this regard. Our anatomically distributed, modular minds are Bundle Theory made flesh. As far as neuropsychology is concerned, Ego Theory doesn't get a look-in. The mental functions underlying our sense of self— feelings, thoughts, memories—are shifting, dynamic processes, scattered through different zones of the brain with no special point of convergence where everything comes together in the form of a stable, ego-shaped *thinking thing*. But we should not be too quick to dismiss Ego Theory. Unlike Bundle Theory, it at least matches our natural intuitions of selfhood. The challenge is to reconcile the intuitions with the actualities of brain function.

Some recent theorizing on the neuropsychology of selfhood cuts usefully across the Ego/Bundle distinction, opening up the possibility of scientific progress. Clinicians and scientists such as Todd Feinberg and Joseph LeDoux have made important contributions, but I am going to focus on the work of the neurologist Antonio Damasio, who is perhaps the foremost neurobiological theorist of selfhood. Damasio distinguishes the *core self* and the *autobiographical self*. The latter is what we generally have in mind when we speak or think of ourselves as unitary, continuous beings with a remembered past and an anticipated future. It's a view of the self tied up with the notion of personal identity. In Damasio's words, the autobiographical self "corresponds to a nontransient collection of unique facts and ways of being which characterize a person." Whereas the autobiographical self is rooted in the past and reaches toward the future, the core self exists only in the present moment, "a transient entity, ceaselessly re-created for each and every object with which the brain interacts." The philosopher Shaun Gallagher has proposed a similar two-dimensional scheme differentiating, in his terms, the Narrative (or Extended) Self and the Minimal Self. Both Damasio's and Gallagher's models of the self can be viewed as descendants of William James's distinction between self as subject and self as object—the "I" and the "Me," but Damasio, more than anyone, has excavated the neurobiological foundations of selfhood. He sees the core self as a product of a dynamic integration of brain systems underlying the perception of internal bodily states ("interoception") along with others engaged in the perception of objects in the external environment ("exteroception"). The network involves interaction between various lower brainstem centers, which in turn communicate with higher cortical centers via intermediate structures in the diencephalon, a set of structures deep in the forebrain. (Don't worry too much about the anatomical details here. You can get the drift of Damasio's ideas without them.)

The core self is the level at which, as Damasio puts it, the brain introduces something into the midst of other mind contents, something that was not present before, "a protagonist," so setting the stage for subjectivity. The core self is thus prerequisite for the autobiographical self, which, in neurobiological terms, resides in higher cortical systems concerned with language and long-term memory. Their job is to construct and continually revise "the story of the self."

So where sits the storyteller? Most likely in the left hemisphere. Through his work with split-brain patients, cognitive neuroscientist Michael Gazzaniga has identified a specialized brain system, located within the left hemisphere, whose job is specifically to correlate unconscious brain processes with happenings in the external world. This provides a framework for the construction of narratives through which the individual makes sense of his or her engagement with the stream of events. He calls this system "the interpreter" and has described its discovery as the most stunning result from his many decades of split-brain research. The interpreter might be understood as the interface between the core self and the autobiographical self insofar as it seems to involve the linkage of in-the-moment cognitive processes with the brain systems involved in story construction.

In Damasio's scheme of things, the systems underlying the core and autobiographical selves are organized hierarchically, with the autobiographical system being entirely dependent upon the core system. He offers epileptic automatism—seizure-induced episodes of automatic, unconscious behavior—as an example of impairment of the core self with concomitant alteration of the autobiographical self. By contrast, there are many examples of the autobiographical self being compromised while the core self remains intact. Acquired memory disorders such as caused by Alzheimer's disease or viral encephalitis radically impair the person's ability to recall past events, to lay down new memories, and to project possible future events. Without the capacity for backward and forward "mental time travel," and the ability to maintain a sense of continuity through the establishment of new memories— hallmarks of the autobiographical self—such patients are to a large extent confined to the present moment, locked in the house of the core self. According to Damasio, then, the core self and autobiographical self systems are dissociable, because it's possible for there to be an intact core self in the presence of a disturbed autobiographical self, but they are not *doubly* dissociable because one does not see the converse—indeed one cannot in principle see the opposite because of the presumed hierarchical relationship of the two systems. I wonder, though, if Cotard's syndrome presents a challenge to this view. Arguably, what we see in the self-negating delusional states associated with Cotard's syndrome (self-negating to the extent that the patient may believe himself to be dead) is an impairment of the core self in the presence of relatively intact functioning of the autobiographical self.

LET'S RETURN, BRIEFLY, to the case of Pat Martino, the recovered amnesic jazz guitarist we met on page 103. One could cite more clear-cut and compelling examples of autobiographical memory loss than the Martino case. My interactions with Pat began many years after his surgery, well beyond the period of the most severe autobiographical memory disturbance, which, from anecdotal accounts, may have lasted some months post-surgery, though at this remove in time, and with no contemporaneous neuropsychological assessments to go by, it's impossible to gauge the rate and extent of recovery with any precision. Superficially at least, some notable idiosyncrasies aside, his autobiographical memory also now seems largely to have recovered. But I think the Pat Martino case remains interesting in the present context, not because it is an especially good example of autobiographical memory disorder— far from it—but because of Pat's repeated claim that ever since his recovery from surgery he has felt himself to be living "entirely in the moment." Although his autobiographical memory systems now seem to be functioning reasonably well, the significance of the autobiographical self, for him, seems to have permanently diminished, while the core self is more sharply in focus. The intensity of his present-moment awareness, and the value he places on this, is a recurrent theme in his conversation, sometimes almost to the point of obsession. In Pat's own idiosyncratic words (taken from his autobiography, *Here and Now!*):

> There's a compression of all the side angles that are within the future and the past . . . after the operation, maybe those two things—dependence on the past and hope in the future—resided

in that part of the brain that was dissected and removed. But whatever was left was "Now"; it had no interest in the future or in the past, and that's where transformation began to take place.

Transformation is his preferred term for *recovery*.

In Cotard's there is a dissolution of the *self of the present moment*, dissolution to the point of experienced nonexistence (a phrase that will have Descartes spinning in his grave). In contrast, Pat Martino claims to live entirely in the here and now. I suppose that's not a bad place to be living if you are a jazz guitar virtuoso celebrated for your extraordinary "in the moment" improvisational skills.

Living the Dream

LONDON, SUNDAY 5 JULY, 2009. LUNCH IS BEING SERVED AT Coq d'Argent, an upmarket French restaurant perched on the roof of a pale-pink and yellow postmodernist building not far from the Bank of England. A young man wearing a Hugo Boss suit buys himself a glass of champagne. Then, glass in hand, he heads over to the roof terrace, where, under the dumb scrutiny of the CCTV cameras, he climbs a barrier and plunges into the dizzying depths of the atrium, arms flung back. Anjool Maldé, a successful stockbroker and entrepreneur, was two days shy of his twenty-fifth birthday. The City of London Coroner's Court recorded a verdict of suicide. It's hard to conceive of an alternative conclusion, given the CCTV evidence. There was no note.

Anjool Maldé's death was front-page news. The papers published pictures of a stylish and good-looking young man gazing confidently at the camera, cocktail glass in hand, surrounded by beautiful women. He had recently bought a penthouse on the Costa del Sol and there were plans for Anjool and his friends to gather there for birthday celebrations. The image projected by the press of a glamorous high-flyer plummeting to a bloody death from an exclusive rooftop restaurant, still clutching his champagne, is as poignant as it is shocking, but it faded soon enough. I had largely forgotten the story when, on Christmas Eve, two years later, I received an email from "a grieving father who lost a special 24-year-old son, an only child." There was no mention of the circumstances of his son's death, and the sender's name, Bharat Maldé, rang no bells. Mr. Maldé, a psychologist, said he was submitting an account of his grief to a journal called *Bereavement Care* and had attached a draft in case I cared to read it. He said the article made reference to Jain, Hindu and Buddhist thinking, which held that, "the best way to deal with suffering is through sufferance," and he thought it chimed with something I'd recently written for *The Psychologist* magazine. I, and others, had been invited to write a few words about a time when psychology had "come to your rescue." I gave it some thought but couldn't come up with anything. So this is what I wrote:

> *Here's a confession. I've been a professional psychologist for 30 years, clinician and academic, but I can't think of a single instance when I've made personal use of my psychological expertise. Even in the darkest times, especially in the darkest times, I never turn to scientific psychology for illumination. I write these words within a few days of the first anniversary of my wife's death, so there have been some very dark days of late. All through, my knowledge of clinical psychology has seemed irrelevant, or if not irrelevant then certainly peripheral to my deepest needs and concerns. This, I know, will sound smug, or disingenuous, or willfully contrarian. But it's true. I am by natural inclination a Stoic. I don't mean in the loose sense of "grimly determined" or "long-suffering," and especially not "stiff upper-lipped." I mean Stoic in the tradition of that broad church of ancient Greek and Roman philosophers— Epictetus and Seneca among them—for whom the question, "How best to live?" was the most important of all. Their collective wisdom boils down to this: negative emotions are a bad thing; banish them through thought and deed. These are the roots of cognitive-behavioral therapy (CBT) of course, the difference being that the Stoics offer an overarching philosophy of life, not just a bag of psychological tricks.*

I thought the "bag of tricks" dig might provoke a bit of discussion but the piece garnered just two brief responses on the magazine's website, one implying I was ignorant of psychotherapy, the other advising, "You just need a friend to go and have a drink with!!!!!"

There I was, declaring that clinical psychology, the discipline I had studied, taught and practiced professionally for over thirty years, was *peripheral to my deepest needs and concerns*. It had never been any use to me. *Even in the darkest times, especially in the darkest times.* I've been fortunate, I suppose, having never suffered serious mental health problems or a brain disorder, but the more general aims of clinical psychology are to reduce psychological distress and to promote well-being. So you'd think my psychological training might have come in handy from time to time during periods of stress, or to lift my spirits at low points, or, taking a broader view, to raise me from my baseline of ordinary well-being toward the peak of full potential. Perhaps, over the years, I had become jaded and turned cynical about psychology, like a heavy-smoking, sedentary, junk-food-stuffing, world-weary cardiologist ignoring his own advice on heart health. But even if I were to reach for the CBT toolkit every time my mood dipped, or launch wholeheartedly into a course of happiness-boosting "positive psychology exercises" to elevate my levels of contentment and self-worth, to what extent would I be addressing those *deepest needs and concerns*? I don't know. What *were* my deepest needs and concerns?

I happened to be strolling through a cemetery when Mr. Maldé's message came through, on a visit to my grandparents' grave, which is something I hadn't done for a very long time. I found a bench and opened the attachment.

It would have been his 25th birthday when two police officers dressed in black brought the news to us that our son Anjool had fallen to his death from an eighth floor swanky City of London restaurant rooftop terrace two days back. He had dressed in his best suit and shirt with matching designer-wear of watch, belt, cufflinks, tie and shoes, walked up to the rooftop restaurant from his nearby flat and had ordered himself a glass of champagne. He had carefully placed the glass somewhere close to the point from which he fell. That's also a detail in which there is meaning.

No one knows what was going through Anjool's mind in his final hours and minutes, those last moments of consciousness. We will never know how, why and when he arrived at the decision to end his life. Quite possibly it was an act of pure impulse. An article published in the *New England Journal of Medicine* the year before Anjool's death notes that between a third and four-fifths of all suicide attempts are impulsive. For around a quarter of people surviving a near-lethal suicide attempt the time between the decision to kill themselves and the actual attempt was less than five minutes. Seventy percent took less than an hour.

My wife's death and Anjool's were very different. Kate's was anticipated. Death was the dogged foe, staved off for years with medical treatments. Anjool's demise was swift and unforeseen. Death promised him immediate refuge. Impulsive and ill-considered it may have been, but the decision to trade his young and healthy life for oblivion was his. He had declared himself to be *Living the Dream* and just a few days later he killed himself. Kate, in her final days, was imploring me to open up to the simple wonder of being alive—*You don't know how precious life is. You think you do, but you don't.* Don't be deluded, she was saying. Wake up from your dream, rub the sleep from your eyes and *see*. They were dissimilar deaths but, in different ways, they bring us right to the heart of the question that was of fundamental concern to the Stoics: How best to live?

"THERE IS BUT ONE truly serious philosophical problem," wrote the French author and philosopher Albert Camus in *The Myth of Sisyphus*, "and that is suicide. Judging whether life is or is not worth living amounts to answering the fundamental question of philosophy." It seems most people consider life to be worth living because the great majority don't choose suicide, or even seriously consider it an option. One might arrive at this judgment after pro-longed contemplation of the pros and cons, or, more likely, give the matter hardly any consideration at all. Life is the default position. Camus himself reasoned (with somewhat convoluted arguments) in favor of life. He died aged forty-six in a road accident. A century before him, the arch pessimist Schopenhauer had a different view. Life was frustrating and meaningless and, on balance, it would have been better not to have been born. But, despite his grim esti-mation of the human condition, suicide was not for Schopenhauer. He died of natural causes at the age of seventy-two sitting on a sofa with his pet cat. Camus said of Schopenhauer that he was "often cited, as a fit subject for laughter, because he praised suicide while seated at a well-set table."

That "fundamental question of philosophy"—*Is life worth living?*—triggers another fundamental question: if it is, then *How to live it?* Most of us don't give that one much serious thought ei-ther. Most of us don't think very seriously about anything. Twenty-first-century Western culture, with its dumbed-down mass media, joyless materialism and mindless obsession with celebrity, gener-ally discourages us from doing so.

IN 2003 *The New Yorker* magazine published an article by Tad Friend, titled "Jumpers," which tells the story of the fatal attraction of San Francisco's Golden Gate Bridge, which, since its opening in 1937, has been the setting for more than 1,600 suicides. Leaping from the Golden Gate Bridge is a fairly reliable way of ending one's life, with something like a 98 percent fatality rate. But there are survivors and, according to Friend, they often regret their decision to leap in mid-air. He quotes Ken Baldwin, who was twenty-eight years old when he leaped from the bridge in 1985. Baldwin decided to jump the railing for fear that standing on the ledge might lead him to lose courage. He counted to ten, but stayed put. He counted to ten again, and this time vaulted over the barrier. "I still see my hands coming off the railing," he told Friend. "I instantly realized that everything in my life that I'd thought was unfixable was totally fixable—except for having just jumped." Another survivor, Kevin Hines, said his first thought as he leaped was, "What the hell did I just do? I don't want to die." In the instant of committing to death, both men preferred life. Baldwin's experience, especially, is remarkable. At the point of no return, he saw solutions to his problems. At first consideration, regretting the decision to leap in mid-air seems an appalling prospect—a last insult to the injuries of a tortured mind—but maybe there's a shred of redemption in it. The last seconds of life are lived in horrified regret, but thereby a sense of life's value has, finally, fleetingly, been restored.

SOME OF ANJOOL'S ashes were scattered on the Thames, with his father chanting the Manglik, a Jain peace prayer. Close relatives placed another portion in the Indian Ocean off Mombasa, and then, on a warm June afternoon a small urn of ashes was lowered from a fishing boat into the Mediterranean Sea off the coast of Spain, near where Anjool had bought a penthouse property just days before his demise. His parents prayed as they lowered the urn into the sparkling sea and, as if to grace the deed, a pod of dolphins circled the boat.

The Myth of Sisyphus

IN GREEK MYTHOLOGY, THE UNIVERSE HAD A FOUR-STORY structure: the sky, the earth and, beneath the earth, the realms of death, subdivided into Hades, the place where humans go when they die, and, way below that, a deep abyss called Tartarus, to which monsters and Titans were banished, and where mortals who had offended the gods suffered fiendish, eternal punishments. Vast distances separated the sky, the earth and the abyss below. According to the poet Hesiod, a bronze anvil dropped from the sky would take nine days to fall to the earth, and another nine days to reach Tartarus.

One of the denizens of Tartarus was Sisyphus, the king of Ephyra (later to be called Corinth). Sisyphus was the craftiest of all tricksters. He had fallen foul of Zeus, the ruler of the gods, by revealing the whereabouts of one of the deity's concubines, Aegina, daughter of the river god Asopus. Zeus had abducted Aegina and confined her to an island (the one that would eventually bear her name). Asopus searched everywhere for his daughter, forlornly, until Sisyphus told him the truth of the matter. When Zeus got wind of the betrayal he zapped Sisyphus with a thunderbolt, but the great trickster cheated death. Thanatos, the god of death, turned up to claim him, bringing along a pair of handcuffs to restrain his prisoner on the journey to the underworld, but Sisyphus tricked him into putting the cuffs on himself, by way of demonstration, and it was Thanatos who became the prisoner. This seriously messed up the natural order of things because, with the god of death locked up in a closet, no one could die. A soldier could be chopped to bits in battle and still report for duty the following morning. This would obviously not do and so, eventually, Ares, the god of war, intervened and Thanatos was released.

Sisyphus was once more summoned to Hades but, again, found a way of cheating death. He told his wife not to bury his body or observe any of the usual funerary rituals and, on arrival at Hades, managed to persuade Persephone, the queen of the dead, that, because he hadn't had a proper funeral she really ought to allow him to return to the world above in order to set matters straight. She fell for it. These gods and goddesses of death were clearly rather gullible. And so, cheating death for a second time, Sisyphus went back to earth and simply got on with his life. Ultimately, though, he received his comeuppance, being dragged back to Hades and condemned to eternal, exquisitely exasperating, torment. He was forced to push a great boulder up a mountain with the aim of tipping it over the other side, but every time he reached the summit the boulder would roll back down again. Over and over again. Endlessly.

We are told nothing about Sisyphus in the underworld beyond the outline of his appalling punishment, but, as Camus said, "Myths are made for the imagination to breathe life into them." And so he does. Camus sees the unbearable toil of Sisyphus in grim detail: the straining body, the screwed-up face, the earth-clotted hands. The agony. And every time, at the summit, the soul-breaking sight of the boulder rolling back down the mountain. "A face that toils so close to stones is already stone itself!" I am quoting from *The Myth of Sisyphus*, the essay in which Camus presents his philosophy of "the absurd." Human life is absurd, he suggests, because of the profound disjunction between our constant search for meaning and the inherent meaninglessness of the godless universe in which we find ourselves, a universe devoid of any eternal truths and values. We can respond, if we choose, by conjuring meaning through a leap of faith, by placing our hopes in a Supreme Being who sets the laws of life. Alternatively, we can face the fact that existence is indeed meaningless, although this raises questions as to whether a life without meaning is worth living. If not, then, says Camus, we are presented with a stark choice. Either we make that leap of religious faith, or we consider suicide.

Camus sees no possibility of reconciliation between the human craving for meaning and the meaninglessness of the universe. The thing to do is to face up squarely to the absurdity of the contradiction. Bear it constantly in mind, and live life to the full, regardless.

So, Camus sees Sisyphus turn and descend to the plain "with a heavy yet measured step," and this, for Camus, is the most interesting phase of the ever-repeating process. This is the "hour of consciousness," the moment when Sisyphus is stronger than his rock. The toil of Sisyphus represents the human condition, ". . . his whole being exerted toward accomplishing nothing." But if he can accept there is no more to life than endless struggle with, ultimately, nothing to show for it, then he can find happiness. Sisyphus is the hero of the Absurd, "as much through his passions as through his torture. His scorn of the gods, his hatred of death, and his passion for life won him that unspeakable penalty . . . This is the price that must be paid for the passions of this earth."

The Problem with the
Problem of Consciousness

KATE AND I WERE OUT ON BEDRUTHAN STEPS BEACH AT LOW tide on a pink summer's evening, strolling among the giant rock stacks. It's a dreamscape, a backdrop for melting clocks and elephants with legs like stilts and, halfway along, we drifted further into the surreal when we came across a multitude of gelatinous sea creatures. Thousands of them, just a few centimeters long, purplish-blue in color, and with a stiff, translucent sail along the back. In the middle of this multitude was a large, dead jellyfish, a jellyfish the size of an umbrella. They had all given up the ghost.

Standing over it, I wondered, because such wondering is what I do, if the jellyfish was any less conscious now, dead and decaying, than it had been in life, pulsing gracefully beneath the waves. Kate rolled her eyes at that, but I told her jellyfish consciousness was something I used to get my students to think about. I expect you did, she said. One of my lectures opened with a set of photographs. There was a picture of a fertilized human egg; a jellyfish; a wasp; a mouse; a thoughtful-looking chimpanzee; and a little girl, about five years old. The question was, which of those living things was conscious? Most people would say the little girl and the chimpanzee, for sure, were conscious, and probably the mouse to some degree, and that the egg certainly wasn't. But the wasp and the jellyfish? *What do you think?* Kate said, and I said I hadn't a clue. Who can say what it's like to be a jellyfish? What is consciousness anyway? No one really knows. And Kate, with a swift *knight's move*, said that maybe consciousness was like a jellyfish.

Back home, I did some research and identified the little purple jellies we'd seen on the beach as *Velella velella,* also known as by-the-wind sailors. They normally live way out on the surface of the ocean, moving around at the mercy of the winds, which, when especially strong, can blow them hundreds of miles across the seas to flounder on the beaches of Europe and America like the vessels of a doomed Lilliputian armada. I thought about Kate's jellyfish theory of consciousness—*You know,* she said, *not a single animal, not a single thing*—and I fell asleep that night with a silly rhyme going round my head. *Velella velella, the jellyfish fella / No arms, no legs, no wings nor propella . . .*

The problem of consciousness, the very *idea* of consciousness, used to drive me nuts. It kept me awake at night. How was it possible for streams of color and sound, and thoughts and dreams, to spring from the soil of the brain, the stuff I'd weighed and sliced and measured in my neuroanatomy classes? I was inclined to agree with the philosopher Colin McGinn, who said the brain is *just the wrong kind of thing to give birth to consciousness.* "You might as well assert that numbers emerge from biscuits or ethics from rhubarb." It was tantalizing and, as McGinn himself put it, "the more we struggle the more tightly we feel trapped in perplexity." But I got over it. I wriggled free of the trap and these days the problem of consciousness hardly bothers me at all. It's not that I found a solution. I don't think there is a solution, at least not for the problem as it's conventionally framed, but I now see consciousness in an entirely different light. If the brain was the wrong kind of thing to give birth to consciousness, I thought, what could possibly be *the right kind of thing*? And it dawned on me that there was no right kind of thing, nothing that could conceivably, in McGinn's terms, *give birth to* consciousness. Consciousness was not a thing to be born. It was not something that the brain could, in any meaningful sense, be said to *produce* or *generate,* as clouds produce rain and dynamos generate electricity. That was a seductive, but false, way of thinking. I'd told Kate I hadn't a clue what consciousness was, but now I feel I do. Squinting hard, I think I glimpse the mystery peeling away and, guess what, I reckon that when you look at it a certain way consciousness really is something like a jellyfish or, rather, something like *Velella velella* the jellyfish fella, who, strictly speaking, isn't a jellyfish, though you might mistake him for one.

I've wondered if I'm kidding myself. Maybe I'm getting old and brain-jaded and just can't be bothered to think about this stuff anymore, not seriously, really just can't be bothered going round in circles all the time, trudging the mire of the mind-body problem, getting nowhere. That's something philosophers are trained to do. I don't have the stamina. So I'm shrugging at the mirror with arched brows and a wistful *who cares?* pressing of the mouth. You win, mate, you *win*. You just carry on tormenting the philosophers. I'll go away and tell myself a *just-so* story because, when it comes to explaining consciousness, everything's a just so story in the end.

There's a grain of truth in all that, but before my own consciousness shuts down forever I would still like to feel that I have seen some sign of an accommodation, a cordial agreement that we can leave one another alone, the mind-body problem and me. And that's where I am. *Entente cordiale.* I'll try as best I can to explain my liberating change of perspective but first we should acknowledge, straight up, that there really is a problem to be solved. It can't be dismissed as a "pseudo problem," and consciousness can't be "explained away," just as the stars of the Milky Way can never be explained away. We might in time arrive at a radically different view of what consciousness is, just as down the centuries people have held different views of what stars are, but both stars and consciousness are facts of the universe. Consciousness is something genuinely in (or about) the universe that calls for explanation.

Here's how we conventionally think of the mind-body problem ("mind-body"; "mind-brain"; "mind-matter"; "problem of consciousness"—I'm using these terms interchangeably). There is a *world out there*, a world of physical objects, forces and events. We flesh-and-bone human beings are part of it. But every second of our waking lives is bathed in a swirl of thoughts, feelings and sensations that occupy a private, inner space, a mental sphere that seems of quite a different character from the physical world of objects and forces. So, we feel obliged to ask, what is the relation of one to the other and, in particular, how does the insentient, physical stuff of the brain—the proteins, the fats, the sugars and the salts, the cells and their multifarious molecules, the 1,200 cubic centimeters of *gloop* that fills our skulls—how does that stuff create awareness? It's beyond dispute that brains are crucial for consciousness but, cubic centimeter by cubic centimeter, it would seem preposterous to suggest that brain matter *itself* is conscious. Taking the brain as a whole, though, consciousness somehow happens. How? That's the problem of consciousness, the great mystery that Nietzsche called *the knot of the world*.

The modern mind-body problem is a legacy of the dualist philosophy of René Descartes. One stormy night in November 1619, the twenty-three-year-old Descartes, at that time a soldier in the Bavarian army, had a series of three powerful dreams. More accurately, I think, it was just two dreams with an odd episode in between which might have been an attack of what neurologists call "exploding head syndrome" (or, more prosaically, "episodic cranial sensory shock"). The dreams had ghosts, a school, a church, a whirlwind that made Descartes spin like a spinning top, a man with (he thought) a melon, a dictionary with missing pages and an anthology of poetry with a missing poem. The first dream, the one with the ghosts and the whirlwind, was terrifying and had Descartes wondering if he'd been visited by a "bad genie." The thought of it kept him awake for two hours. When eventually he dropped off again he was almost immediately roused by what he perceived as a loud bang and he opened his eyes to see the room full of sparks. This was (perhaps) the exploding head bit. The last dream, about the books, was the most powerfully affecting. He first noticed the dictionary, which he thought would be useful, and then an anthology of poetry appeared, *Corpus Poetarum*. Opening it, he saw the line, *Quod vitae sectabor iter?* "What path in life shall I follow?" A stranger appears and gives him another poem that opens with the words, *Est et non*, "Yes and no" (or sometimes translated as "What is and what is not"). It continues with Descartes searching for, and failing to find, the poem in the anthology. The dictionary reappears, this time slimmer, having lost some pages. The symbolism of the dreams has been analyzed by a number of scholars, including Freud, who partially agreed with Descartes' own interpretation. "What path in life shall I follow?" clearly indicates that he is at a crossroads in his life, and "Yes and no" tells him his mission is to find ways of distinguishing the true and false in all human knowledge (represented by the dictionary). Freud also saw the melon as a sign of sexual frustration and thought Descartes would have been well advised to find a mistress. Descartes, equally eccentrically, but significantly for the history of science and philosophy, saw it as a symbol of the importance of meditation and solitude. But, at any rate, he sets about his mission, divinely inspired, he believes, to discover infallible methods for securing knowledge of the world.

Plenty of people before Descartes had thought of body and soul as distinct entities, going back to Plato and beyond, and all of the major religions promote the idea of the separateness of bodies and souls, but Descartes' absolute demarcation of the territories of mind and matter set the terms of debate for the next three centuries. He came to the conclusion that the universe consisted of two fundamentally different kinds of substance, *res extensa* (material substance) and *res cogitans* (mental substance). The body is a machine, a purely physical thing, inhabited and controlled by a nonmaterial soul, which is a "thinking thing," in our terms a *conscious mind*. In this context mind and soul are equivalent. The soul/mind is the essential, inner core of personal being. It lies beyond the reach of objective analysis in a different order of reality than the material world. This account of things has come to be called *Cartesian dualism* (from Cartesius, the Latinized version of the philosopher's name), or substance dualism, and is pretty much in line with our natural intuitions. We (our minds) decide on a particular course of action and our bodies are directed to act.

The question is, how can a non-material soul interact with a material body in order to control it? We know how neurons (physical things) interact with muscles (physical things), and Descartes himself developed a physiological theory of the reflex arc, but how can the material and the non-material interact? As far as I, and most neuroscientists are concerned, Descartes had no satisfactory answer. He envisioned "animal spirits" flowing in the cavities of the brain and through the nerves to the muscles, and, on the grounds that it was in the middle of the brain and nobody had a clue what it did, speculated that the pineal gland was the point of connection between mental and physical substance. But still, *how* did they interact? Three centuries later the Oxford philosopher Gilbert Ryle dismissed Cartesian dualism as the myth of the ghost in the machine. Despite the widespread rejection of substance dualism, the Cartesian way of carving up the world has left us with some problematic dichotomies: mind versus matter; inner versus outer; subjective versus objective; observer versus observed. You could say they only bolster our natural intuitions, but natural intuitions sometimes get in the way of scientific inquiry. The Sun does not, most obviously, orbit the Earth even if it—equally obviously—appears to do so.

The alternative to dualism is *monism*, the view that the world consists of just one kind of stuff, be it mental or material. Most neuroscientists are in the materialist camp and think that, ultimately, the mind, consciousness included, can be explained in terms of the physical operations of the brain. This is what's known as a *reductionist* approach, the mental being reduced to the physical. If we could fully explain the workings of the brain the problem of consciousness would simply dissolve away, because brain activity and consciousness would be understood to be one and the same. Francis Crick, co-discoverer of the structure of DNA and,

in his later years, consciousness researcher, encapsulated this reductionist assumption in what he called the *Astonishing Hypothesis*: "that 'you,' your joys and your sorrows, your memories and
your ambitions, your sense of personal identity and free will, are
in fact no more than the behavior of a vast assembly of nerve cells
and their associated molecules." The challenge is to identify the
neural correlates of consciousness, precisely which features of brain
function are necessary and sufficient for consciousness to come
about. But Crick's hypothesis seems to rest on the assumption that
brain activity *is* indeed sufficient for consciousness, beyond being
just necessary (most people accept that brains are necessary). According to the philosopher Alva Noë, this is not so much an assumption as an *ideology* about what we are. It's taken for granted
that consciousness is somehow located within the brain, that the
brain contains something that thinks and feels, that is, essentially,
us. Ghostly stuff aside, this is a view that isn't so far removed
from Descartes. He took the internal thinking thing to be a non-
material "substance"—a soul—whereas modern neuroscientists
identify it with the stuff of the brain. They have taken a materialist
turn, but in one crucial respect the idea is basically the same: consciousness is located in the head. Crick's Astonishing Hypothesis,
by the way, is not all that novel, having been around in one form
or another at least since the seventeenth century, when the likes
of Thomas Hobbes, a staunch critic of Descartes, and Baruch Spi-
noza were arguing for the material basis of the soul. What is really
astonishing about Crick's Astonishing Hypothesis, says Noë, is
how astonishing it isn't.

Another view is that, even with a complete knowledge of brain
function, there would still be an explanatory gap between physical
levels of description (the activity of neurons in the brain) and mental life (experiencing the beauty of birdsong, the sadness of loss,
etc.). There is physical stuff on the one hand and conscious experience on the other, and they call for different forms of description
and analysis. This is the position taken by the philosopher David

Chalmers. He contrasts the Hard Problem and the Easy Problems of consciousness. The Easy Problems cover the bread-and-butter questions of cognitive neuroscience. How are psychological functions (perception, memory, emotion, language, and so on) distributed across different brain systems? How are they implemented at the nitty-gritty neuronal level? Note that these questions are "easy" in the same way that the problems of cosmology or particle physics are easy, which is to say not very easy at all. The point is they can be studied scientifically. Yet how are we to account for conscious awareness from the private, first-person perspective of actual experience, as opposed to public, third-person observations of brain and behavior? What is the redness of a red balloon? Is it something more than the neuronal response of eye and brain to electromagnetic radiation of wavelength 740 nanometers? Or take the example of pain. We can analyze pain biologically and behaviorally as a state caused by bodily damage that predisposes the individual to avoid further damage. Beyond this we can investigate how pain is realized in the nervous system in terms of the activation of so-called A-fiber and C-fiber nerve pathways. But there is nothing here about the feelings involved. That's the Hard Problem, accounting for the feelings. It's hard precisely because of the shift from an objective to a subjective point of view. Objective science can tell us about the functions of pain and the associated patterns of physiological activity, but still it seems reasonable to ask, why does it *hurt*?

Suppose I got hold of a futuristic brainscope that could provide me with detailed information about the activity of your brain. It supplies data not only at the systems level, just as currently available brain scanners do, showing which regions of the brain "light up" when you engage in different kinds of activity, but also at the fine-grained neuronal level. I can tell precisely which brain cells are active and interacting. You agree to wear the scope for a day or two as you go about your usual business. You wear it like a hat. I check in midmorning and the data streaming onto my screen inform me that you are drinking coffee and eating chocolate. The scope has detected this by decoding neuronal activity in the thalamus, a structure embedded deep within the center of your brain, and from activity in the primary gustatory (taste) cortex, an area enfolded within the parietal lobe. It also takes into account activities in other, widely distributed brain regions, including those involved in memory, mood, vision and motor control. The brainscope informs me that it's white chocolate that you are eating and I ask it which brand. The chocolate bar is out of view so visual readings are unhelpful, but with a quick scan of the visual memory centers it comes back with the information that it's Green & Black's Organic. You are, apparently, enjoying it very much according to readings from the limbic lobe and the cingulate gyrus. In just a few seconds I know precisely the pattern of brain activity involved in your enjoying a cup of coffee and a chunk of Green & Black's Organic white chocolate, down to the level of individual brain cells. But *so what*? I've solved the "easy problem" of identifying brain patterns associated with a particular person enjoying a particular brand of chocolate on a particular occasion, but it gets me no nearer to grasping your experience of the texture and flavor of the chocolate or the pleasure you get from eating it. I can only get that secondhand and inferentially, from the exercise of my own imagination and in this case my imagination, lets me down because I'm not so keen on white chocolate myself. I can conjure up the texture and flavor, but not the intense pleasure. I could buzz around your brain with my brainscope and discover all sorts of things, including a lot of stuff you don't know about (last night's forgotten dreams, for example), and plenty of other information you might prefer me not to know—in fact, stuff you'd prefer *yourself* not to know—but as far as eating chocolate is concerned I may as well just ask you how you're enjoying it. Your words will get me as close to the flow of your experiences as the scope does. And, despite the incredibly detailed neuronal information I'm getting, and the close correlations with your reported experiences, I still can't explain how the neuronal activity translates to consciousness.

Back in the eighteenth century, but with futuristic imagination, the philosopher Gottfried Leibniz pictured "a machine whose construction would enable it to think, to sense, and to have perception." He then imagined enlarging the machine "so that one could enter into it, just like into a windmill." And what does he find in the middle of his mind-making machine? ". . . only parts pushing one another, and never anything by which to explain a perception." These days we wouldn't be picturing the levers and cogs of a windmill, we would pitch ourselves into the switches and circuits of a computer, but the same principles hold. In either case it's the implicit suggestion of likewise scaling up the brain that captures the imagination. And we can also imagine scaling ourselves down. Shrink to the size of a microbe and picture yourself in among the neurons, weaving through the cell bodies and the axons and the dendrites. There's silent, invisible electrical activity going on all around you, but not much sign of that, and no sign whatsoever of color or sound, or *anything by which to explain a perception*, let alone a self or a soul.

Part of the problem is that we don't have anything like a coherent, consensus view of what it is we're trying to explain. Consciousness? Well, it's *just obvious* what consciousness is, isn't it? If you favor life over death, if you enjoy sex and sunsets, if you recoil from the smell of excrement and opt for anesthesia when having abdominal surgery, then you know well enough what consciousness is. But turning from ordinary intuitions to face the requirements of science and philosophy, and the challenge of capturing something we might identify for further scrutiny, we find that consciousness stubbornly resists definition.

A pragmatic, commonsense way to characterize consciousness is simply to equate it with being awake. If you can interact with the world, communicate with others, and monitor your own thoughts and actions, then it's safe to say you are conscious. The opposite applies for states such as sleep, coma, anesthesia and death. There are, of course, gradations of wakefulness, and, in clinical practice, neurologists are stocked with well-defined terms for labeling levels of impaired arousal (in descending order of alertness: *clouding of consciousness; lethargy; obtundation; stupor; coma*). Clinicians also know, in broad terms, how consciousness relates to brain function, loss of consciousness being associated with extensive damage to the cerebral hemispheres across both sides of the brain, or damage to a deeper structure called the thalamus, or damage to systems way down in the brainstem. The overlying gray matter of the cerebral cortex, the stuff serving the "higher order" functions of perception, language and memory, plays little part in regulating levels of consciousness.

So far, so straightforward, but equating consciousness with wakefulness won't do. When we sleep we dream. We are not displaying the wakeful, behavioral signs of consciousness but while dreaming we are, for sure, having conscious experiences. The converse, wakefulness without awareness, is a defining feature of the so-called vegetative state, which results from extensive damage to the cerebral hemispheres and thalamus, but with the brainstem arousal systems still functional. Patients go through a regular sleep–wakefulness cycle and retain basic reflexes, such as blinking to a sudden loud noise, but in their eyes-open waking state they show no sign of any awareness of themselves or what's going on around them.

Equating consciousness with wakefulness is problematic enough but *awareness* is the real troublemaker. Awareness, or sentience, that subjective *feeling like something* dimension of conscious experience is much more difficult to pin down. It seems to draw us irresistibly into those stubborn intuitions of the Cartesian *thinking thing*, in its private, impenetrable inner realm.

Whatever it is, consciousness seems fundamental to the value of life, and, come to think of it, maybe that's not a bad starting point if we're trying to work toward a definition. Consciousness: *the capacity of mind without which one's life has no value.* Suppose you are forced to choose between another ten years of normal, conscious life or another fifty of zombiehood, your body going through the motions of life, and appearing as normal to others, but completely devoid of conscious experience. I'm guessing you'd take the first option. Ten years, ten weeks, ten minutes—anything but zombiehood. Life without consciousness would be no life at all. And, fanciful thought experiments aside, consciousness is a key consideration in those sad, real-life medical situations involving decisions over whether to discontinue artificial life support. The body's vital functions can be prolonged pretty much indefinitely with modern medical technology, but it's widely accepted that life-sustaining treatments may justifiably be withdrawn in cases of brain injury so severe as to render the patient permanently un-conscious and beyond hope of recovery. Arguably, because con-sciousness has irretrievably gone, they have already ceased to exist as a person.

We can agree (I hope) that we are not zombies and that it feels like something to be human and alive as opposed to being dead or permanently unconscious, and as opposed to being a ball-bearing or a bicycle, and we can agree that this *feeling like something* is an important (some would say defining) feature of consciousness. Furthermore, we can agree that different experiences have dif-ferent qualities. The taste of apple pie is quite unlike the pain of a headache or the thrill of a rollercoaster ride. But, and this is a crucial point, I think, it's a mistake to imagine that we can dis-til from those different experiences something that is the essence of consciousness, something they all have in common to infuse them with the *raw feels* of awareness, something that makes them *feel like something.* This is a trap set by our intuitions. We need to stop thinking of consciousness as some mystical, ethereal, qualia-drenched, redness-of-red, hurtingness-of-pain, essence of aware-ness. And that's hard because this is such a deeply ingrained way of thinking. So, it's not so much an Einstein of neuroscience we need to figure out a solution to the problem of consciousness as a Harry Houdini, to show us how to escape from the straitjacket of our intuitions about what consciousness is. As Wittgenstein would have it, we need to show the fly the way out of the fly bottle. Philosophy, for Wittgenstein, was more about correcting misunderstandings than solving puzzles, and the philosopher's treatment of a question was like the treatment of an illness. Our view of the world is clouded with fundamental misconceptions, and the job of philosophy is to restore clarity of vision.

I don't have a solution to the mind-brain problem but I have found a way of thinking about it that makes it less tantalizing, less frustratingly hopeless to wrestle with. If Wittgenstein is right about philosophy as therapy then I have hit upon a treatment that works for me. It doesn't involve any startlingly original insights, at least none that I've ever had. I've just taken stock of other people's ideas, turned them over, stuck them together this way and that and built myself a conceptual deck chair on which to lounge and watch the world go by without being bothered to distraction by the mind-brain problem. This shift is not so much an intellectual step forward as a sense of achieving an intuitive grasp, a visceral understanding, even, of ideas I could already see the logic of.

In one way or another my therapy is all about shaking off Cartesian habits of mind. The first thing is to ditch those qualia. In 1966 a distinguished Austrian philosopher, Herbert Feigl, was giving a lecture on the problem of consciousness. His celebrated Vienna Circle colleague Rudolf Carnap was in the audience. Feigl had written an influential book advocating a materialist view of the mind-brain problem, in which he argued that the mind was fundamentally a physical entity. But yet his thesis had a nagging incompleteness. Qualia, he conceded, those raw feels of awareness, remained a scientific mystery. "But, Herbert," Carnap interjected, "you are failing to take account of the alpha-factor." Science was, in fact, now making good progress toward explaining raw feels.

This was news to Feigl and he was eager to hear more about the alpha-factor. "Very well," said Carnap. "You tell me what a raw feel is and I'll tell you about the alpha-factor."

Yes, *you tell me*, and *you tell me* what the alpha-factor might be, you tell me what consciousness is for that matter, if you think qualia are a central part of the story. I live next to a sports field, along which runs a row of white poplars. The trees reach close to my window and, right now, on a summer's day, the view is a trembling mosaic of sage-green leaves and gray-white branches. It's lovely. I look through the window. I *see* the trees. I *respond* to them. What does it add to say that I am conscious of the scene? Only confusion. And how many qualia am I having when I look at even so simple an object as an apple (the shine and the shadow, the tints) let alone when I stare into those dancing leaves with their myriad shifts of light and shade? I am *conscious of* the trees and qualia are an intrinsic feature of this experience? No, let's just say *I see the trees. I respond to them.* To go beyond this is to summon those ectoplasmic qualia into the room, and this is a seminar not a séance.

At all costs avoid thinking of consciousness in monolithic terms as a uniform, all-pervasive "thing in common" between different experiences, a kind of irreducible mental substance. Picturing consciousness in this way is akin to the way nineteenth-century physicists imagined all of space to be filled with an invisible, rarefied, elastic substance they called the ether, a hypothetical medium through which light and other electromagnetic waves were thought to travel. The ether does not exist. Modern physics did away with it. Consciousness as a kind of ethereal substance (invisible, rarefied, elastic, held in common by all experiences) does not exist either. We don't need "qualia" to explain "experience," just as we don't need the ether to explain the propagation of light waves. What is it that infuses different experiences with the raw feels of awareness, with that something that makes them *feel* like something? Nothing does. The obsession with qualia, like the tunnel-vision focus on internal brain processes, is something that keeps a lot of thinkers shackled to the Cartesian cart. We really have to ditch Descartes.

A difficult part of my therapeutic process has been the letting go of the idea that consciousness is *all or nothing*. I read somewhere a lovely phrase: *The distance between silence and a whisper is infinite*. I don't remember where I came across it, and I've tried to track it down to no avail, but I recall that the writer was making a connection with consciousness. He/she was suggesting that the distance between insentience and the merest glimmer of consciousness was also infinite. Consciousness is all or nothing. You are either conscious or you are not, just as you are pregnant or not. There are grades of consciousness, as there are stages of pregnancy, but the drop from that merest flicker of awareness to oblivious insentience is an infinite abyss. At some stage in the evolution of life from the bacteria of the Precambrian era to the brainy birds and mammals of the present day, organisms became conscious. At some stage in the progression from conception to adulthood individual organisms (some of them, but certainly us humans) become conscious. Out of Darkness Cometh Light! Perhaps it's just a photon, but a single photon is *something rather than nothing*.

No! Resist the *all or nothing* temptation. It lures us into the substance trap, the idea that consciousness is a *thing*, if only a very tiny thing, like a photon, or a fluid, invisible thing like a breeze, or a completely nonmaterial thing like a Cartesian soul. On the evolution of consciousness, philosopher and scientist Daniel Dennett says the idea of drawing a line on the complexity scale of life between animals who are and who are not conscious is "profoundly pre-Darwinian; there might be all manner of variations from poet to possum . . . to protozoon, and no 'essence' of consciousness to discover." Indeed. There is no essence. *Stop searching for the essence!*

It's been a list of *don'ts* and *ditchings* so far, primarily the legacies of Descartes. I've been focusing on the habits of mind I think we need to break, but we're still in the heart of the labyrinth. I'm not afraid because I think I see the Minotaur of the mind-body problem for what it is—a mythical beast, not a real monster. But we still need to find our way out, and the Ariadne's thread of thought I think we need to follow will lead us from the intuitive sense of consciousness as input (perception) to the less intuitive idea of consciousness as output (action). That's the way out.

Dennett writes dismissively of the "Cartesian Theater," the idea that there is a location in the brain that somehow contains the contents of consciousness, parading, as it were, across the stage. This, naturally, implies a viewer, and the metaphor fits with our intuitions that we are spectators as much as participants in the flow of conscious experiences. The neurologist Antonio Damasio is fond of movie screen metaphors. The general drift is that consciousness is to be understood as, fundamentally, an act of perception or, better perhaps, the perception of perceptions. But some philosophers and psychologists have turned to thinking about consciousness in relation to embodiment and action, rather than, as is traditionally, intuitively, the case, in relation to processes of perception and thought going on within the brain. There have been many contributors to this movement. The late Susan Hurley made a seminal contribution with her book, *Consciousness in Action.* Other influential figures include the philosophers Andy Clark, Alva Noë, and Kevin O'Regan. Max Velmans is in the same ballpark, too, I think, or at least in the vicinity, and Nicholas Humphrey, whose ideas we visit later, has plotted an evolutionary story about the behaviors leading to the development of consciousness. Some of that lot will object to being lumped together as a "movement." There are some sharp disagreements among them, and maybe they are a movement only in my head insofar as they've got me to look at consciousness with a fresh eye.

Hurley refers to "the classical sandwich model of the mind" with its clear separation of perception (input from world to mind) and action (output from mind to world) with cognition crammed between. We should instead think of mind and consciousness in terms of a dynamical relation between brain, body and world, a set of processes built into patterns of perception and action, without specific location. Noë suggests that rather than think of conscious-ness as an internal process, like digestion, we should consider it as an activity, like dancing. Trying to explain consciousness by focus-ing exclusively on neuronal activity is like trying to explain what dancing's all about by studying only the muscles. The dancer's muscles are crucial, but only part of the picture. Likewise for neu-rons and consciousness. Experience is not something that happens in us, Noë says; it's something we do. Experience itself is a kind of dance, "a dynamic of involvement and engagement with the world around us."

Consciousness is something we do. Absorbing that concep-tual flip from input to output was the turning point in my problem of consciousness therapy. Consciousness is not some mysterious thought-bubble stuff floating in and around our heads. Conscious-ness is the doing, the acting on the world, the whole brain-body-world dynamic. And if you remind me that bees and ants and jellyfish are embroiled in brain-body-world interaction, well so they are. (Jellyfish don't actually have a brain, although they do have rudimentary sensory nerves capable of detecting light, tem-perature, touch, salinity and so on.) And does their dynamic en-meshment in the world make these creatures conscious? I don't know. I haven't a clue! It's an interesting question that might (or might not) be amenable to empirical research within a revised, non-Cartesian, *actionist* paradigm. But if we persist in looking at consciousness through Cartesian lenses, seeing it as a thing, an enclosed, internal process, a light that switches on, and all that, we'll get nowhere.

Maybe it would help if we just abandoned the word conscious-ness altogether for the purposes of understanding mental life. With a shift of perspective, a turn to embodied cognition and the extension of mind through body into world, perhaps consciousness will go the way of *élan vital*, the life force, or essence, once thought necessary to distinguish living things from inanimate matter. Living things, we now understand, have no such added ingredient. The mystery of *élan vital* eventually just faded away. So too will the mystery of consciousness if we can escape our flawed ways of thinking about it.

Since we returned to the subject of jellyfish just now, let me tell you I thought there was something to be said for Kate's Jellyfish Theory of Consciousness—*You know, not a single animal, not a single thing.* It was based on a misconception because jellyfish are, in fact, single animals, single things. She'd lumped them together with the *siphonophores,* a quite separate class of jellyfish-like marine creatures. Unlike a true jellyfish, a siphonophore is not a single organism, though it looks and behaves like one. It consists of a colony of smaller organisms working in concert to present the appearance of a larger, singular animal. The stranded sailors we encountered at Bedruthan, *Velella velella,* are siphonophores, but the best-known example is the venomous Portuguese man o' war, which is more often than not mistaken as a jellyfish. But let's not get bogged down in the taxonomy of gelatinous sea creatures. The analogy's the point. This: just as *Velella velella* and the Portuguese man o' war are collections of individual creatures functioning as a larger animal, so too consciousness is constructed from a multitude of independent processes, sensory and cognitive, yet presents (is *experienced*) as a unified field of awareness, something over and above its component parts. In each case "the larger animal" is both an illusion and *not* an illusion. Despite appearances, the siphonophore isn't "really" a singular animal but it truly *functions* as one, eating, excreting, reproducing, generally making its living in the world, and, at the species level, subject to the forces of evolution just like any other species. Consciousness, likewise, is dual aspect: at one level a colony of autonomous subsystems, each with its own tight focus on this or that facet of the world, and, at another level, a coherent, larger *something,* a system, a scheme of things, that recruits those subordinate systems so as to register, react to and reflect upon events in the world, *as a whole.* But the trap, the very root of the problem of consciousness, is to think of that larger something as, indeed, a singular *thing*, rather than a scheme of things or a process, or a network of relations—brain, body, world—in constant flux. Consciousness as a creature of the fourth dimension: time.

Coda

In his book *The Demon-Haunted World*, a paean to science and skepticism, Carl Sagan writes of a yearning to communicate with his dead parents. He misses them terribly and longs to believe that something of them still exists, some essence, some element of personality. What wouldn't he give for even five or ten minutes a year, to tell them about their grandchildren, to share his news, to give them his love? Irrationally, a part of him wonders how they are. "Is everything all right?" I feel the same about Kate. How I would love to tell her about the grandchildren, the ones she knew, the one she didn't, the one currently under construction. I see her smiling face in my mind's eye. The boys? You'd be proud. Me? Fine. I still get those . . . I call them *stabs of absence*. But, you know. She rests a hand on mine, gives it a squeeze. Like Sagan with his parents, a part of me wonders how she is. He knew, and I accept, that nothing remains, and yet the sense that *something does* never goes away. It's a fundamental, *nuclear*, part of how we relate to people even when they no longer exist.

Sagan didn't believe in an afterlife and nor do I, but, in line with the teachings of the major religions, most people do. It is widely presumed that at death the soul transfers from the body to a non-physical realm of heavenly bliss, or somewhere less desirable, for all eternity. Or, alternatively, having left one body it returns to the earthly realm to occupy another, human or animal. These are supernatural beliefs and you might think that any survival of consciousness and personality after death necessarily involves the supernatural, but Douglas Hofstadter, a distinguished cognitive scientist, speculates otherwise. He is able to conceive of a survival of personal consciousness after death that is entirely natural.

The son of a Nobel Prize–winning physicist, Hofstadter had achieved his own considerable success as an author and scientist with research interests focusing on the philosophy of mind, and especially the cognitive underpinnings of the sense of self. His book, *Gödel, Escher, Bach* ("GEB"), tagged "a metaphorical fugue on minds and machines," won the 1980 Pulitzer Prize for general nonfiction. Then, in December 1993, tragedy struck. His beloved wife, Carol, died quite suddenly and unexpectedly of a brain tumor. What hit him the hardest, he says, was not his own personal loss, which was considerable, but, rather, Carol's: the sudden obliteration of her consciousness, and the loss of a future life so rich in promise. Struggling to come to terms with it all he turned to his close friend and colleague, Daniel Dennett, with whom some years earlier he had co-edited a book called *The Mind's I* (which, incidentally, Kate had given me as a Christmas present when it first came out, thereby galvanizing my own interests in the neuropsychology of selfhood). They communicated by email because Hofstadter was on sabbatical in Italy at the time and Dennett was back in Massachusetts. Hofstadter offers an edited version of their conversation (predominantly his side of it) in his book *I Am a Strange Loop*, which, elaborating on lines of thought first presented in GEB, offers an account of the patterns of brain activity underlying the sense of self. This is what preoccupied him, the thought that Carol's consciousness had not been completely obliterated, that her mind, or at least some elements of it, had found habitation in his own brain. Were their selves interfused such that she, in some real sense, was seeing through his eyes? Thinking about Hofstadter's conjectures, I was walking through the park on a sunny day, my shadow sharp and full on the path in front of me. Then, passing beneath trees, it dissolves into dancing leaves, but still there are recognizable fragments of my form. Perhaps the postmortem remnants of another person's consciousness are something like that.

Perhaps, but if so I think they would amount to nothing more than simulation—my brain simulating Kate's joy, for example, in the presence of our grandchildren. I feel something of the way I know Kate would have felt, but, still, that's my consciousness, not hers. It's a long step to imagine an independent, autonomous consciousness, specifically Kate's consciousness, cohabiting my brain, seeing through my eyes, hearing through my ears with me unaware of the particulars of her experience. If something of Kate's personal consciousness continued to exist in my brain after her death, then presumably it would have existed in similar form when she was still alive, in which case, if it truly were her consciousness she would have been aware of its contents. Walking down a country lane, taking in the sights, the scents and the sounds, she would, albeit to a lesser degree, also have been aware of the sights, scents and sounds surrounding me at a train station a hundred miles away.

Hofstadter's bereavement was more catastrophic than mine. His wife's death was sudden and unexpected, whereas I had years to prepare for Kate's. Their children were small. Ours were grown. And Kate had enjoyed a good ten years more life than Carol. Perhaps in a similar situation I would be clinging to hopes of my wife's survival, in however attenuated a form. But would I wish there to be wisps of Kate's consciousness wafting through my brain, seeing through my eyes? (I'm sorry, I'm slipping back into Cartesian mode here.) No, I don't think I would. I adored Kate for what she was, a loving, caring, sensuous, fun-loving, flesh-and-blood woman whose laughter and tears intermingled with my own. I cherish her memory, but perish the thought of a ghostly, looking-wanly-through-my-eyes version of her drifting through the dreary suburbs of my gray matter. I think of her every day. But you have to let go.

Tyger, Tyger

PICTURE A FOREST. IT'S THE DEAD OF NIGHT. A STORM IS raging. At the heart of the forest a brawny oak sways and shakes in the howling wind. Resplendent with a million leaves (invisible in the dark), its branches whip this way and that, clashing with the boughs of neighboring trees. Oak, sycamore, beech and alder bustle shoulder to shoulder. Billions of raindrops are bombarding millions of shuddering leaves. Rainwater drips and trickles through the canopy, cascading down to the protected understorey of saplings and shrubs, dogwood and holly, into the undergrowth where ants and beetles scurry, down into the earth. All is action and connection. Countless drops of water, countless twigs and leaves in constant, fleeting interaction; gusts and zephyrs shaping large- and small-scale patterns of association and energy flow. Picture yourself sheltering under the oak tree, in the dark.

And then, out of nowhere, there flashes the image of a tiger. It's visible for two or three seconds. But not to you. All you see are black shadows and glimpses of gloomy cloud over the canopy of the trees from time to time. The rain stops; the wind subsides; the forest is still. Have moonlight if you want. There's no sign of a tiger. Why would there be in an English forest?

Tyger Tyger, burning bright
In the forests of the night . . .

A Devil's-advocate thought. Suppose consciousness boils down to patterns of brain activity. Well, if so, why just brain activity?

Why not patterns of anything? There's nothing special about neurons. They're just physical things doing physical things. It's the patterns that matter, not the stuff making the patterns. Now what if those patterns get replicated in some other physical system? The pattern of brain activity that produces the taste of vanilla, or the color blue, or the sound of a trumpet, gets replicated once in a while, purely by chance, in the pattern of the interaction of ocean waves, or ears of corn in a rippling cornfield, or grains of sand in a sandstorm, or stars in a galaxy. Hardly likely? But just think of the vast numbers of water droplets in a rainstorm or leaves and branches in a forest. If consciousness is no more than patterns of interaction between material stuff, then you might have flashes of disembodied consciousness happening all over the place. Maybe that's what your conscious experience comes down to right here and now—a fleeting pattern of interaction between wind and rain and trees in a wild storm.

Wriggling Redly

NICHOLAS HUMPHREY'S CONTRIBUTIONS TO THEORETICAL and experimental psychology are wide-ranging. Among other things, he discovered the phenomenon of "blindsight" (more on which below). He has studied mountain gorillas in Rwanda, and has made pioneering contributions to the psychology of social intelligence and its role in evolution. But for some time now his prime focus has been the evolution of consciousness.

Humphrey's central contention is that consciousness is grounded in bodily sensation and action rather than in perception and thought. Put simply, we don't so much *have* sensations, as *do* them. Sensation is, as he puts it, on the production side of the mind rather than the reception side. When we look at a red poppy we are "redding." The evolutionary history of sensory enactments like redding (or salting, hotting, and so on) can be traced to the bodily reactions of primitive organisms responding to different environmental stimuli, noxious and nutritive. Imagine an amoeba-like animal floating in the ancient seas. Like all other organisms it has a structural boundary, which is the frontier between "self" and "other." The animal's survival depends on cross-border exchanges of material, energy and information and, as it moves around, some events at the border are going to be "good" for it and some "bad." It must have the ability to respond appropriately, "reacting to this stimulus with an ouch! To that with a whoopee!" as Humphrey puts it. At first the responses are localized to the site of stimulation, but evolution endows more specialized sensory zones—this for chemicals, that for light—and a central control system, a protobrain, which allows for coordinated responses to specific stimuli. So, when, for example, salt arrives at its skin, the animal detects it and makes a characteristic response, a wriggle of activity. It wriggles "saltily." Red light elicits a different kind of wriggle—the animal wriggles "redly." These primitive reactions are the prototypes of human sensation.

With the march of evolutionary history life gets more complex for the animal and it becomes advantageous for it to have an inner representation of events happening at the surface of its body. One way of accomplishing this is to plug into those systems already in place for identifying and reacting to stimulation. The animal's representation of "what's going on?" (and what it "feels" about it) is achieved by monitoring what it is doing about it. "Thus . . . to sense the presence of salt at a certain location the animal monitors its own command signals for wriggling saltily . . . to sense the presence of red light, it monitors its signals for wriggling redly." Such self-monitoring by the subject of its own responses is the prototype of "feeling sensation."

Evolution then takes the animal to another level at which it comes to care about the world beyond the frontier of its body, so that, for example, it becomes sensitive to chemical and air pressure signals of the proximity of predators or prey. This requires quite another style of information processing. "When the question is 'What is happening to me?' the answer that is wanted is qualitative, present-tense, transient, and subjective. When the question is 'What is happening out there in the world?' the answer that is wanted is quantitative, analytical, permanent, and objective." The old sensory channels continue to provide a body-centered picture of what the stimulation is doing to the animal, but a second system is set up to provide a more objective, abstract, body-independent representation of the outside world. This is the prototype of perception. At this stage the animal is still responding to stimulation with overt bodily activity but eventually it achieves a degree of independence and is no longer bound by rigid stimulus-response rules. It still needs to know what's going on in the world, so the old sensory systems stay in service, and it still learns about what is happening to it by monitoring the command signals for its own responses. But now it has the capacity to issue virtual commands that don't result in overt action. In other words, it no longer wriggles. Rather than going all the way out to the surface of the body, the commands are short-circuited, reaching only to a point on the incoming sensory pathway. Over evolutionary time the target of the command retreats further from the periphery until "the whole process becomes closed off from the outside world in an internal loop within the brain." Sensory activity has become "privatized."

According to Humphrey, all perception is unconscious. Sensation and perception are separable, he argues, and it is the sensory systems, not the perceptual, that underlie conscious awareness. He points to the neuropsychological evidence. As a Cambridge PhD student in the late 1960s, working under the supervision of Larry Weiskrantz, he made some astute observations of a monkey whose primary visual cortex had been surgically removed. The monkey, called "Helen," should have been completely blind. But Humphrey developed a hunch that she could somehow "see." He found time to sit by her cage and play with her and it soon became clear that she was watching him. Over the next seven years he became her friend and tutor, taking her for daily walks in the woods and fields, encouraging her and coaxing her, "trying in every way to help her realize what she might be capable of. And slowly but surely her sight got better." Watching her run around the room, negotiating obstacles and picking up crumbs from the floor, you would assume she had normal vision. This could not be the case and, with another leap of intuition, Humphrey sensed that she still didn't really *believe* that she could see. He noticed that sometimes, when she was afraid or upset, for example, her "sight" would fail her and she would stumble around as if in the dark.

Helen, of course, couldn't say what she saw, or thought, or believed. But within a few years "blindsight" (as it came to be called) had been identified in a human being. Weiskrantz, encouraged by

Humphrey's observations, was conducting experimental studies of a patient, "DB," who had undergone surgery to remove a tumorous malformation of blood vessels in his right occipital lobe. The operation required extensive removal of the primary visual cortex, resulting in a large scotoma (blind spot) in the left visual field. Predictably, DB was oblivious to stimuli presented in his blind field. Yet although he reported no visual sensation he could, when prompted to guess, accurately report the position and shape of the objects presented. If you were to present a red light spot in the blind field of a blindsight patient and ask him what he can see he would claim to see nothing. In all likelihood, though, he would be able to judge that the spot was red. He would thus be perceiving redness but not experiencing it. In Humphrey's terms, he would not be "redding."

This brings us to the heart of the matter, indeed the heart of mattering—the *matteringness* of consciousness. Consciousness *matters*, and we should, if we follow Humphrey's line, be close to grasping just *what* the matter is. In blindsight, the person "sees" but has no sensation of seeing. The "raw feels"—the qualia—of the shape and color of the seen object are missing. His brain is informed of objects in his visual field but "he" (the experiencing person) is disengaged from them. They are of no matter to him. According to Humphrey, the function of sensation (what it *does*) is to track the subject's personal interaction with the external world, thereby creating that sense we have of being present and engaged, "lending a hereness, a nowness, a me-ness, to the experience of the present moment." Those raw feels are mattering-in-action and they bring into being the "us" to which things matter. As Gottlob Frege put it, "An experience is impossible without an experiencer. The inner world presupposes the person whose inner world it is."

Selfhood and consciousness are thus entwined "in-the-moment," and Humphrey has explained the role of sensation in both, but has he nailed what Daniel Dennett has skeptically termed "Factor X," that mysterious flame of phenomenal experience, those sparkling qualia that light up the "explanatory gap?" Well, says Humphrey, if the elusive Factor X has to do with anything at all, it has to do with time. Phenomenal consciousness is about the temporal "depth" of the present moment. The subjective "now" is, paradoxically, extended in time: it is "temporally thick." We experience it not as an infinitely thin sliver of time but as a moment in which times present, past and future overlap. We travel through life as if in a "time ship," which "has a prow and stern and room inside for us to move around." The problem is that the notion of the "extended present" is fundamentally incoherent to the commonsense mind. Our experience ("the thick moment"—an amalgam of past, present and future) is at odds with our understanding of the linearity of time. We can't get our heads around those ineffable qualities of consciousness because, as the philosopher Natika Newton points out, the very nature of Factor X makes it "analytically, ostensively and comparatively indefinable." According to Humphrey, it is precisely this that gives consciousness its mysterious, out-of-this-world qualities, and creates the irresistible intuition of mind-body duality. Nature has performed a stupendous conjuring trick: the illusion of the soul. It is an illusion that at once creates and valorizes us as conscious entities. It is thereby an adaptive illusion. Consciousness matters, says Humphrey, because its function is to matter. It has evolved to create in human beings a Self whose value is intrinsic and whose life, therefore, is worth pursuing. Even beyond death. But like the Wizard of Oz, the brain is an illusionist, and the seemingly magical, ethereal quality of consciousness is an illusion. *Pay no attention to that man behind the curtain!*

The mystery of consciousness and the riddle of selfhood are thus different facets of the same enigma, and if we see things this way, then, I would suggest, the mind-matter problem starts to lose its menace. The mysterious, magical aura of consciousness will never entirely fade because a sense of the ethereal and otherworldly are right at the heart of what consciousness is, or rather what it does. It's the brain's way of creating a certain, special model of itself—projecting its "self" and other conscious "selves" on an illusory mental plane that appears to intersect with the physical world. And, by the way, I am *not* saying that "consciousness is an illusion," which would be circular and silly because illusions can only occur *in* consciousness, in other words they presuppose a conscious experiencer. Nor am I saying, "consciousness doesn't exist," which is even sillier. It's just not what we tend to think it is.

When it comes to reconciling brain function and consciousness it may not so much be an explanatory gap that we have to contend with as, in Thomas Metzinger's phrase, an intelligibility gap. Metzinger makes the point that we could have a satisfactory theory of consciousness but if the theory is not intuitively plausible we will not "experience the truth" of it. That's dead right, it seems to me. As Humphrey himself asks, if Factor X fell into our lap would we even realize it? And that's what's so hard about the hard problem of consciousness. We may one day solve the problem in scientific terms but still not fit the solution into our intuitive frames of imagination.

Still, whatever insights and solutions are on offer, consciousness will never be "explained away." Rest assured, you can carry on enjoying sex and sunsets.

Immortality

I ONCE RECEIVED AN INVITATION TO CONTRIBUTE TO A COLloquium on the Law of Futuristic Persons. A futuristic person, apparently, is "a being who claims to have the rights and obligations of a human but who may be beyond currently accepted notions of legal personhood." You are a futuristic person if, for example, your mind runs on a computer rather than a brain, or if you have been brought back to life after a period of legal death spent, say, in the deep chill of a cryonic vault. The (at present very remote) prospect of having your conscious mind uploaded into a computer may not be so enticing, but who wouldn't choose to extend their flesh-and-blood life by the fruits of biomedical science? We do so already, of course, taking the life-sustaining gifts of modern medicine for granted. Given the chance of a little more life, and yet a little more, most of us would take it, eking out our lives indefinitely. We'd keep on keeping on. And radically increased longevity is no longer a fantasy. Quite likely, as the present century unfolds, advances in genetic engineering, nanotechnology and regenerative medicine will deliver on their life-extending promise, at least for a rich and privileged few. According to the Cambridge gerontologist Aubrey de Grey, a "cure" for human aging is just around the corner. Periodic repairs to the bodily machinery using methods already within the scope of science—stem cell treatments, gene therapies and the like—have the potential to halt the aging process, in effect to reverse it. De Grey believes that the first person to reach a thousand years of age has probably already been born. Onwards to immortality. But would I really want to live forever, or even for a thousand years? Oh, I think I'd give it a go.

One argument against is that a life stretching interminably into the future would become terminally boring. There are only so many places to visit, so many symphonies to hear, so many lovers to embrace. Yes, so many, so many! Life is infinitely rich. The possibilities for new knowledge and experience are endless. So I don't buy the boredom argument. Another argument doubts the durability of human identity. It goes like this. I may opt for immortality, but as the millennia roll out I would become vastly altered, physically and mentally. Constantly reengineered, my body would change shape, color and proportion according to fashion, and my brain capacity would be boosted a thousandfold by neural implants. I would no longer be "me." The quest for immortality is thus, literally, self-defeating. But is this so different to the changes we undergo in the course of an ordinary lifetime? Our bodily tissues are continually renewed; cells die and are replaced. The average age of the cells that make up an adult human body (that is, your "true" physical age) is around seven to ten years. Our physical and mental structures are in constant flux. Only patterns survive, the DNA and the tumbleweed bundles of proteins and fats, sugars and salts, flesh and fleeting impressions, rolling down the days and years. Patterns, that's all.

Resurrection

I SEEM TO BE MAKING A SPEECH TO AN AUDIENCE OF STRANG-
ers, fifty or sixty, standing, drinks charged, with huge windows
behind them, and through the windows a blazing urban night-
scape. I clear my throat. I raise my glass.

"On this very special day, my one-hundred-and-fiftieth birthday, it's good to be surrounded by those I love."

The words are producing themselves. That's what normally
happens, I suppose, but I am more than usually aware of the disparity between my thoughts (*Who are these people?*) and the sounds
issuing from my lips.

Fiftieth birthday? No, it was definitely one-hundred-and-fiftieth.
I remember my fiftieth well enough. I was on a transatlantic flight
to New York. When we left Heathrow I was forty-nine years old.
I turned fifty mid-Atlantic but, entering Eastern Standard Time
was, briefly, forty-nine again. Later on I checked my watch and
said to my traveling companion, "Just turned fifty again," and we
laughed and had a weary conversation about turning fifty. It was a
big one for him. Once is bad enough, he said. Twice is taking the
piss. When you're thirty you say, God, I'm getting old and everybody laughs. Same when you're forty. Then you're fifty, and it's not
funny anymore. You just *are*. But numbers didn't bother me like
that.

Here comes Selene, and the century-old memory skitters off
into the gloom of unlit neurons. *A hundred years!* Selene not yet
born. Her mother not yet met. *You haven't got a daughter*, says a little whisper in my head, but with Selene's embrace I feel the warm

wash of a thousand memories and I know who I am, and I know who they are. And, for sure, I know who she is.

"There's no denying I feel old, but in body, not spirit. Or do I mean the other way around?"

Laughter.

"And there I go, slipping back into the old ways of thinking: mind and body, spirit and substance. There's no excuse. The ghost in the machine . . . well, you know."

The ghost in the machine was exorcised long ago. Selene is the living proof. Selene, my darling, uploaded daughter. Let's see. Chronologically, Selene is ninety years old. Physically, she's a genetically reengineered woman of thirty. Psychologically, well, these days you have to keep an open mind about psychological ways of being. But one never stops worrying about one's children, and the uploading, the transfer of information from old brain to new, was, I confess, a little troubling. I have always felt a measure of responsibility for the current popularity of mind transposition. My writings helped create a climate of acceptance. There was a time, long ago, when we had bodies and souls, or imagined we did. Or imagined we didn't. A time when life was simple. It was nasty, brutish, illusory, deluded, difficult and short, but *simple*. This is how it went: you were born; stuff happened; you died. It was a time of simple certainties. Your soul, for sure, drifted on someplace else. Or, for sure, there was no such thing as the soul and, whatever there was, the "I," the "You," the "Me," the introspective self, the personality, was doused like a flame, for all eternity. That was my message. I wasn't the only one peddling it and I never laid claim to a single original thought on the subject. I can't take all the blame.

Selene's kiss was settling. I recognize them all now. Sons and daughters, their partners, their children, their grandchildren, their grandchildren's children. Ky, my beloved, raises her glass and smiles her lemon-slice smile. Before Ky there was Lily, and before Lily there was Mira, and before Mira there was Fay and before Fay there was Kate.

———∞∞∞———

SELENE TAKES MY hand. Her voice is soft and urgent. "Dad," she says, "you're nervous. You of all people! But, believe me, you will not regret this. *Will not.* I promise." This gives me a churning in the gut. Love and apprehension.

What was that I once wrote? *The laser beams of neuroscience are beginning to penetrate the philosophical fog of centuries.* What crap! The real philosophical fog was just beginning to roll in. But that wasn't how it felt at the time. Mind science was coming of age. Traditional methods of studying the brain, my methods, were giving way to neuroimaging and computational neuroscience and, by 2030, the micro-architectures of cognitive function were rapidly unfolding. The inventor and futurist Roy Richter predicted that the brain would be fully reverse-engineered by 2040, with hardware and software available for the implementation of human intelligence in a non-biological substrate. He was wrong, but not far wrong. Richter intended to live forever. He just had to get to a hundred, or thereabouts. According to another of his predictions, the Transcendence, aka the Age of Technological Hyperintelligence, would dawn in the mid-2050s. So he followed rigorous dietary and physical training regimes, he took bucketfuls of supplements every day to live as long as possible; to boldly go, to stride, to stroll, to creep, to crawl to get, by hook or by crook, to the bright rising sun of the Transcendence, and therefore immortality. He didn't make it. His demise was spectacular and has never been fully explained. Some consider it to be beyond rational explanation. But that's another story. Death, anyway, became irrelevant. I'm sure I wasn't the only one suppressing a wry smile the day Roy Richter died and, Almighty Zeus, what a preposterous way to go! Struck by lightning while out jogging on his ninety-ninth birthday, a thunderbolt from a clear, blue sky, straight and true as an assassin's bullet, just hours before his scheduled mind upload. There were fantastical reports of a golden chariot hurtling through the heavens, but the lightning strike was real enough. Richter would have been the first human being to reach the cyberplains of immortality, but his apotheosis was not to be.

I digress. The technology inevitably became entangled with the philosophy. As those post-millennial neuroscientists marveled at the sparkling, dare I say *spectral*, patterns cascading from their high-resolution brain scanners, they were nagged by a mischievous question: *Who's running the show?* How does the brain pull off the personal identity trick? "Soul" didn't figure in the lexicon of neuroscience, but its secular cousin, "Self," was stirring, declaring its chimeric presence in the neural zoo with a leonine growl and a flick of the serpentine tail.

"Dad, you don't have to." Selene's head is resting on my shoulder, as it did when she was three years old, but now it's she who's giving comfort and reassurance. "We can call it off and just carry on with the party."

Selene is all the proof I need. I can't let her down. Can I?

Selene takes my hand and leads me to the chamber where my gift awaits. I see my reengineered body, which sits motionless; the limp corpse of a young man prepared for resurrection. Its carbon nanotube brain circuitry lies dormant but will soon be infused with my digital ghost. Like Selene, I chose thirty. That was a good age. Unlike her, I resisted the temptation to tinker with cosmetic details. Take me or leave me. And I'm opting for a conservative, Level 1 transposition. My new brain will run, like the old one, as a stand-alone unit with unenhanced software. Selene is Level 3—enhanced and hive-mind compatible. She is fully immersible—and these days mostly immersed—in the Web of Awareness, aka *The Hive*. "What's it like?" I ask her. "Inconceivable," she says, her eyes mocking my nostalgia for puny individualism. Then she tells me that the time has come. I sit in the chair adjacent to the corpse, wishing that it didn't have to be quite so ceremonial. I stare at the palms of my hands, deep into the ridges of the skin. *I'm still here.*

What was that old movie? *The Ruling Class?* Peter O'Toole, the psychotic aristocrat. "When did I realize I was God?" he says. "Well, I was praying and I suddenly realized I was talking to myself." My epiphany was less grandiose. It was quite the opposite. I realized I was talking to myself but no one was listening.

Happy Birthday, Dad.

Goodbye, Selene.

A Shower of Gold

AFTER A YEAR OR SO, IN OUR FINAL UNDERGRADUATE YEAR at Sheffield, Kate and I got a room together in a shared house. It had a splendid view over the woodlands of the Porter valley, and a double bed. I played Bob Dylan a lot. Kate preferred Leonard Cohen. There were posters on the walls. One of mine was *Earthrise*, the picture taken by the astronaut William Anders while orbiting the Moon on the Apollo 8 mission, the one with the Earth, a swirl of cloud and gray-blue ocean, hanging over the barren, gray surface of the Moon. One of Kate's, Blu-tacked to the wall at the head of the bed, was a reproduction of Gustav Klimt's *Danaë*. A beautiful young woman lies coiled, eyes closed, lips parted, cheeks flushed. Her diaphanous robe has fallen aside. Golden-red tresses cascade across bare shoulder and breast. Her knees are raised to expose a voluptuous expanse of thigh and a rolled-down stocking hangs at her ankle. A torrent of liquid gold gushes from above and plays between her legs. Her right hand seems poised to clench. When I see the picture now I also see Kate, coiled naked beside me, Sunday morning sunlight streaming through the window.

In taking Danaë and the shower of gold as his theme, Klimt was following in the footsteps of a distinguished line of painters, including Titian and Rembrandt. Here's the story that inspired the art. Acrisius, king of Argos, longed for a son and heir but had only a daughter, Danaë. He consulted the Oracle at Delphi and was told that he was fated to have no sons. Worse, Danaë would give birth to a boy and this grandson was destined to kill him. So Acrisius had built an underground chamber of bronze in which to imprison his daughter and thus prevent her from having contact with men.

In later versions of the story her prison is a tower of bronze. Either way, the prison was impregnable, with just a narrow aperture for the passage of food and drink and the flow of fresh air. Zeus felt pity and passion for Danaë and, although the cell was impenetrable to mortal men, it could not thwart the desire of the king of the gods. A shower of gold poured through the slit and into Danaë. This was Zeus. In due course a baby boy was born. Refusing to believe the story of the boy's divine paternity, Acrisius locked mother and son in a wooden chest and set them adrift on the sea. But Zeus saw them safely to the shores of the island of Seriphos where a fisherman, finding the casket caught in his nets, took the pair into his home. As the son of Zeus, the boy was destined for the life of a hero, and the hero the boy would become was none other than Perseus, who would slay Medusa, the snake-haired Gorgon, and rescue the princess Andromeda from Cetus the sea monster. As the Oracle foretold, in due course he did indeed kill his grandfather, although quite unintentionally. The old man was accidentally struck by a discus thrown by Perseus in a sporting contest. Zeus chuckled. *No, no, nothing to do with me, honest.*

DANAË AND THE shower of gold is a story of magic and myth. As ways of making sense of the world, magical and mythical thinking can be contrasted with mechanistic thinking, which is characteristic of the rational, observational approach that science takes. Then there is mentalistic thinking, which we intuitively employ in making sense of our own and other people's thoughts and actions, in terms of feelings, desires, beliefs and intentions. And so we have the Four Ms: mechanistic; mentalistic; mythic; and magical. They can be placed on a continuum from the *natural* to the *supernatural*.

Mechanistic explanations of the world fall entirely within the natural realm, and magical explanations fall entirely within the supernatural. Mentalistic and mythical explanations to some extent straddle the natural and the supernatural. By natural I mean *obeying the laws of nature* and by supernatural I mean *defying the laws of nature*. Now, philosophers of science have spent entire careers trying to clarify the meaning of *laws of nature* but the gist of an explanation will suffice for now. So, for example, gravity is the force that causes objects to fall toward the ground. It is a fundamental and universal force through which masses are attracted to one another—an apple to the Earth, the Earth to the Sun—and its properties can be described with mathematical precision. If I were to fall from the top of the Empire State Building the 82.6-kilogram mass of my body would, like any material object, hurtle in the direction of the center of the Earth, accelerating at the rate of 9.8 meters per second squared, and would hit the ground in under 10 seconds. In so doing it would be obeying the laws of nature and, in particular, the laws of motion and gravity. If, however, it started to decelerate halfway down, affording me a gentle touchdown on West 34th Street, then the laws of nature would have been defied and so it could be considered a supernatural event. Perhaps God had intervened, or perhaps in the mental frenzy of my descent I had discovered within me some extraordinary powers of "mind over matter." Either way it would be a miracle.

Not everyone would be convinced. For the naturalist there is no division between the natural and the supernatural because there is no such thing as the supernatural. Everything in the universe is subject to the laws of nature and supernatural phenomena are simply beyond the bounds of possibility. Naturalism, by the way, has nothing to do with *naturism*. "If you don't believe in anything supernatural," says the philosopher Tom Clark, "—gods, ghosts, immaterial souls and spirits—then you subscribe to naturalism, the idea that nature is all there is." Naturists, on the other hand, are people who like playing ping-pong in the nude.

Trying to explain my decelerating fall, witnessed by hundreds, would be a mighty challenge for the naturalists, but they would be bound to search for a scientific explanation. Perhaps I'd been caught up in some freak meteorological event. A small whirlwind had whipped up, taken me in its grip and borne me gently down. Perhaps, for reasons unknown, I'd been saved by the invisible intervention of an extraterrestrial intelligence using technical powers beyond our current comprehension. Or perhaps the event simply didn't happen. I imagined it, or everyone in the street imagined it: an illusion or a mass delusion. A true naturalist could never be convinced of the supernatural fact of the matter. Naturalism is ultimately an act of faith, and true believers, like the true believers of any creed, are incorrigible.

Suppose one day I realized I had the ability, like Zeus, to take the form of different animals or to shower the woman of my dreams with gold. I tell my naturalist friend I can do these things. She thinks I am speaking metaphorically but can't figure out what I'm getting at. No, I tell her. I can *really* do such things. So now, perhaps she thinks I am joking. She waits for the punch line. It's not a joke, I insist. I really, literally, can take the shape of a snake or a bull if I so choose, or turn myself to liquid gold and rain from the sky. Go on then, she challenges. Prove it. OK. Animal or shower of gold? She goes for the latter and, with a wave of my hand, I disappear and droplets of gold start falling from the sky. Soon my friend is standing in a glistening pool of gold. She looks dumbfounded. Then a knowing look crosses her face. I'm dreaming! Possibly, I say, with disembodied voice. Or hallucinating. Now she's looking worried. Or, I've hypnotized you. But you're not, and I haven't. OK, she says. I'm impressed! How did you do it? The last drop of my golden self rolls down her cheek. Supernatural power, I tell her, but she's having none of that. She never would. Anything, *anything*, but the idea of contravening the laws of nature. Nature is all. As Sherlock Holmes said, when you have excluded the impossible, whatever remains, however improbable, must be the truth. And, for naturalists, the supernatural is impossible.

Returning from a conversation with his father's ghost, Hamlet says to his skeptical friend, "There are more things in heaven and earth, Horatio, than are dreamt of in your philosophy." Beyond the screen of earthly life, he means, there lies a world of unearthly supernatural phenomena, a world of spirits and spooks. But for the great biologist and science communicator J. B. S. Haldane the line provokes a different turn of mind. "My own suspicion," he writes, "is that the universe is not only queerer than we suppose, but queerer than we can suppose ... I suspect that there are more things in heaven and earth that are dreamed of, or can be dreamed of, in any philosophy." In other words, the natural world enfolds mysteries beyond human intellect and imagination. This, he says, is the reason he has no philosophy and it's his excuse for dreaming. So if it's unfathomable, awe-inspiring mystery you crave there's no need to invoke the supernatural. Nature is enough. But people have always invoked the supernatural, no less in this modern scientific age, and I mean not just that large numbers of people continue to believe in gods and ghosts, but that even naturalists are prone to supernatural turns of mind. Magical thinking is a part of being human. I dare say even Richard Dawkins has a pair of lucky underpants.

The Rape of the Moon

"I T'S A TWELVE OH TWO."

Neil Armstrong and Buzz Aldrin stand shoulder to shoulder in the flimsy lunar module, the *Eagle*, as it hurtles in silence across the face of the Moon. Altitude: 50,000 feet. Speed 3,000 mph, more than five times faster than a 747. To say that the men are standing side by side is somewhat arbitrary. They are in the weightlessness of space and, from their perspective, lie horizontal relative to the surface of the Moon: horizontal, face down, and flying backwards, feet first. Armstrong rotates the spacecraft 180 degrees so that now they are staring out into space with the Moon at their back. The maneuver gives the landing radar clear sight of the lunar surface so as to establish their precise altitude, but the alarm has been triggered. There's a problem with the on-board guidance computer.

"Program alarm. It's a twelve oh two," says Armstrong.

"Twelve oh two," Aldrin confirms.

They don't know what alarm code 1202 means, and nor does the flight director, Gene Kranz, back at Mission Control in Houston. Kranz consults Steve Bales, engineer and flight controller, who in turn consults a twenty-four-year-old computer specialist, Jack Garman. Garman understands that code 1202 signals an "executive overflow." The guidance computer is not keeping pace with its workload. Time is ebbing. What next? Silence. Static. Vital seconds slip by with no information forthcoming from Houston. Armstrong is unflustered, as ever, but his tone is insistent: "Give us a reading on the twelve oh two program alarm." The Moon's surface rolls beneath them, a mile a second.

———— ✖✖✖ ————

The Moon is the Earth's sole natural satellite. It was formed four and a half billion years ago when the Earth collided with another planet. Some material from the mantle and the core of the two planets would have merged and other debris would have been jetted into orbit, eventually coalescing by force of gravity to form the Moon. This is the currently favored theory. When we look up at the full Moon we see a disk roughly the size of the United States of America. The Moon has a diameter of 2,160 miles, a circumference of 6,790 miles, and it weighs 81 quintillion tons. A quintillion is 1 followed by eighteen zeros: 1,000,000,000,000,000,000. That's a million trillion. I have no grasp of the difference between a million trillion and a billion trillion. The Moon orbits the Earth at an average distance of 240,000 miles and the length of a lunar day is 27.3 earth days. A lunar eclipse occurs when the Earth passes between the Sun and the Moon. Solar eclipses occur when the Moon passes between the Earth and the Sun. In a total eclipse the Moon fully and precisely obscures the disk of the Sun. The occlusion is precise because, viewed from the surface of the Earth, the Sun and Moon are the same apparent size. This is due to a remarkable coincidence of distance and size, the Sun being 400 times more distant than the Moon but having a diameter 400 times greater. These are the material, mechanical facts of the matter, but there are other ways of looking at the Moon.

"It's OK." Garman, the young computer specialist, thinks they need not abort the mission as long as the alarm is intermittent, not continuous. He thinks it's OK. He *thinks*. But the final decision rests with Bales and, if anything goes wrong, that's his responsibility too. He decides in an instant. He must. So the message comes through from Houston: "Roger. We got it. We're go on that alarm." In other words, ignore it. It has cost 24 billion dollars to get to this point in the Apollo Moon program and the lives of three astronauts. Two more lives are now in jeopardy, but the quick-fire decision to continue with the mission regardless of the program alarm rests on the opinion of a twenty-four-year-old backroom boy who thinks it's OK. The risk is deemed "acceptable." Whatever that means. Armstrong reorients the *Eagle* once more, this time toward the vertical for the final stages of the descent, but the alarm continues, becomes more insistent. Communication with Mission Control begins to fracture.

Alexander von Humboldt, the great nineteenth-century naturalist and explorer, wrote of an ancient belief that the Moon was a huge celestial mirror. He traced the belief to the Greek poet Agesianax but says it was widely held among the people of Asia Minor even into his own day. He was once astonished to hear a very well-educated Persian remark that the mirror hypothesis was accepted by many of his fellow countrymen: ". . . 'what we see in the Moon,' said the Persian, 'is ourselves; it is the map of our earth.'" It's a beautiful thought. Look close enough and we would see ourselves in true context: small specks of life on a lonely globe suspended in black space. First-person perspective flipped to third. Here becoming there. Practitioners of magic turn the tables. They capture the image of the Moon in a mirror specially washed with oil and mugwort, and kept wrapped until the Moon is full. Then, having caught the Moon's reflection, they stare into the glass to catch prophetic visions.

THE STRESSED GUIDANCE computer has made minor navigational errors that could have major consequences. The rapidly descending lunar module is several miles askew of its anticipated position and headed for a large crater surrounded by an even larger area strewn with rocks and boulders, impact with which could damage the landing gear or slash holes in the flimsy outer fabric of the

202 A THOUSAND RED BUTTERFLIES

vessel. Armstrong knows they are close to the point of no return, the "dead man's box," at which a decision must be made either to continue with the landing, perhaps at risk of fatal hazard, or to abort. As commander of the mission it's his call. What next? He takes control, slowing the descent, surging forward and, at less than 1,000 feet, skimming and skirting the obstacles below. To the team at Mission Control he seems to be taking an age. Fuel is running low. He doesn't say much.

ALDRIN: "Watch your shadow out there . . . fifty, down at two-and-a-half, nineteen forward . . . two hundred feet, four-and-a-half down . . . five-and-a-half down . . . one hundred twenty feet . . . one hundred feet . . . OK. Seventy-five feet."
HOUSTON: "Sixty seconds."
ALDRIN: "Forty feet, down two-and-a-half. Kicking up some dust."
HOUSTON: "Thirty seconds."
ALDRIN: "Contact light. OK. Engine stop."
HOUSTON: "We copy you down *Eagle*."
ARMSTRONG: "The *Eagle* has landed."

As Armstrong and Aldrin were orbiting the Moon, preparing for their descent to the lunar surface, I got into a fight. It was on a riverbank in Wales. I was hanging out with some lads after a week of Cadet Force training exercises and hard, under-age drinking at the British Army training camp at Brecon, where the real, grown-up soldiers of the Parachute Regiment looked down on us with benign disdain. I recall a warm, summer's afternoon and a cheerful atmosphere. We were heading home. Someone cracked a joke and, totally out of the blue, my friend Danny punched me in the face. Then he punched me again. He was bigger than me, a lock forward to my outside half, and he could swing a punch, but I retaliated with such blind ferocity that, within seconds, Danny was on his knees, his face a chaos of blood and snot. The violence felt wrong and right at the same time. Either way, I suppose it was an under-standable reaction for a fifteen-year-old being punched in the face. I have no idea why he hit me. *"What was that for, Danny? What was that for?"* He didn't offer an explanation and we never discussed it. Ever. Not so many years later, Danny got married and I was his best man, but we lost touch over time. Kate didn't much like him, she said, but had always rather fancied him. Recalling the fight, I wondered what she remembered of that day of the Moon landing. It was the day her periods started, she said.

THE MOON'S BIG silver disk is a projection screen for the imagi-nation. Our pattern-detecting brains, predisposed by folklore and tradition, have found all manner of things on the Moon's surface: an elephant; a hare; a four-eyed jaguar; a woman weaving; a man bearing a burden of sticks. Most plainly, though, we see the Man in the Moon. Plutarch, the first-century historian and biographer, wrote a long essay, *On the Face Which Appears in the Orb of the Moon*, concerning the Moon's size and substance and its role in the world and the life of the soul. We are intimate with the Moon. It symbolizes the passage of time, life and death. With its wax-ing and waning we see a projection of our own mortality. Each month, out of nowhere, a new Moon is born. It grows, it flourishes. It diminishes and dies. The poet Shelley saw the Moon as a dying woman "lean and pale,/Who totters forth, wrapp'd in a gauzy veil,/Out of her chamber, led by the insane/And feeble wanderings of her fading brain."

THERE ARE SYSTEM checks to run through and the astronauts are not due to step down onto the lunar surface for another seven hours. According to the schedule, they are expected to rest, eat and sleep before going outside. Aldrin's thoughts turn to the spiri-tual. He has brought with him a small silver chalice, a vial of wine and a little plastic package containing a wafer of bread. He is preparing to perform the Christian ritual of Communion. In the one-sixth gravity of the Moon the wine curls slowly and gracefully into the chalice. He reads a quotation from the Bible, John 15:5. *I am the vine, you are the branches. Whoever remains in me, and I in him, will bear much fruit; for you can do nothing without me.* He eats the bread and drinks the wine, representing the body and the blood of Christ. Armstrong has declined an invitation to participate in the ritual. He looks on respectfully, or at least that's Aldrin's interpretation.

IN GREEK MYTHOLOGY the Moon, Selene, fell in love with a beautiful shepherd, Endymion, as her beams fell on his sleeping face. Dreading the prospect that, like all mortals, he would age and die she got Zeus to grant him eternal youth, which he did. But also eternal sleep.

AT 02:56 COORDINATED Universal Time on 21 July 1969, Neil Armstrong becomes the first human being to set foot on the surface of the Moon. He has prepared a few words to mark the occasion: "That's one small step for (a) man, one giant leap for mankind." The "a" is in parenthesis because, although it is required for the sentence to make any sense at all, it was not uttered by the astronaut—or, at least, it was not heard by the millions of earthlings tuned in for the Moon landing. Armstrong later claimed to have voiced the "a" and that, somehow, the crucial little word was lost in transmission. We'll never know. Who cares? Aldrin descended the lunar module ladder for his Moonwalk about twenty minutes later. I wonder if thoughts of his mother crossed his mind on his lunar stroll. She had killed herself the previous year in a state of depression triggered, apparently, by the prospect of her son's forthcoming Moon mission and the fame and the acclaim it would be bound to bring. She feared she wouldn't cope. Her maiden name was Moon.

THOSE FIRST FOOTFALLS in the moon dust punctured a membrane—the thin, porous film that separates reality and imagination. Some saw this as an act of destruction, a violation. It was a sentiment that inspired Tom Stoppard to write *Jumpers*, a play about philosophy, murder and Moon landings. He was curious to know whether, "if and when men landed on the Moon, something interesting would occur in the human psyche." He cites a statement from the Union of Persian Storytellers ("if you can imagine such a thing") to the effect that a Moon landing would be damaging to the livelihood of the storytellers. The Moon as romantic metaphor, as symbol of love and dreams, of the unconscious mind, the passage of time, of life and death, would be diminished. The veil would drop and the Moon would be revealed as rock and dust. We knew that anyway, but there are different ways of knowing.

Then it happened. Armstrong and Aldrin walked on the Moon and nothing changed.

The View from the Bottom of the Well

THE SCIENCE UNDERLYING THE APOLLO MOON MISSIONS, indeed, all science, is rooted in the rich intellectual soil of ancient Greece during the period between the sixth and fourth centuries BCE, when *logos* (reason) was gaining ascendancy over *mythos* (mythological stories) and the gods were gathering doubters, or at least downplayers. The origins can be traced to Thales of Miletus and his quest to explain natural phenomena without recourse to mythology and superstition. The exquisite execution of the Moon landings owes most to Isaac Newton's formulation of the laws of motion and universal gravity in the seventeenth century, but the ancient Greeks laid the conceptual foundations, establishing patterns of thought that led ultimately to Newton's profound insights, and it all started with Thales of Miletus. Thales is credited with being the first person to predict a total eclipse of the sun: that which occurred on 28 May 585 BCE. According to Herodotus it happened in the middle of a battle between the Medes, an ancient Indo-Iranian people, and the Lydians from Anatolia (modern-day Turkey). Thales' home city of Miletus lay just to the west of Lydia in the coastal region of Ionia. Herodotus describes the event thus:

> . . . *a battle took place in which the armies had already engaged when day was suddenly turned into night. This change from daylight to darkness had been foretold to the Ionians by Thales of Miletus, who fixed the date for it in the year in which it did, in fact, take place. Both Lydians and Medes broke off the engagement when they saw this darkening of the day: they were more anxious than*

they had been to conclude peace, and a reconciliation was brought about . . .

It's not known how Thales managed to predict the eclipse, if, indeed, he really did. He left no writings and we know of his ideas and achievements only through the works of later writers, in particular Aristotle. He traveled widely, including to Egypt where he learned the geometry that became the paradigm for his deductive reasoning. Among other accomplishments he determined the diameter of the Sun and the Moon, and identified the relationships between the solstices, the seasons and the sun's varying trajectory across the sky through the course of a year. He recognized the advantages of navigating by the constellation Ursa Minor rather than Ursa Major, which had been the convention. The advantage is conferred by the tighter orbit of Ursa Minor about the pole star, and therefore its less variable location in the sky. Such knowledge was of practical concern to the Milesian mariners who were significant maritime traders. Aside from his achievements in philosophy and cosmology Thales was himself a highly successful merchant, who was quite ready to apply his scientific nous to the world of business. It was weather patterns and not the gods that determined the success of a harvest, he realized, and weather conditions were to some degree predictable. Appeals for divine intervention were useless, even to Zeus "the Cloud-Gatherer," Zeus "the Thunderer." Observation was the thing; and prediction based on observation. Accurately predicting a high yield of olives one year Thales bought up all available olive presses in the area and leased them out at substantial profit to meet the increased demand.

But Thales was an exception. Most Greek minds at that time were still richly infused with the gods and heroes of Homer's great epic poems, the *Iliad* and the *Odyssey*, composed some two hundred years earlier but drawing on much older mythological traditions. Likewise, in the *Theogony* ("birth of the gods"). Homer's near contemporary Hesiod had offered a compendious and authoritative account of the origins of the universe, from the yawning void of *Chaos* via the Titans (the divine offspring of Earth and Sky) to the establishment of the Olympic Pantheon with Zeus at its head. This was the received Reality, the mythological Truth. The disorder of *Chaos* had been supplanted by the order of *Cosmos*. It was an order established and overseen by capricious deities. Thales wanted none of that. He dispensed with the gods.

If, as is often claimed, Thales was the first scientist, then perhaps he was also the first absent-minded scientist. There is the story of him gazing up at the stars so intently that he fails to watch where he is walking and falls into a well. It's just possible, though, that Thales entered the well more by design than by accident. As Pliny the Elder later observed, "The sun's radiance makes the fixed stars invisible in daytime, although they are shining as much as in the night, which becomes manifest at a solar eclipse and also when the star is reflected in a very deep well." If Thales knew that stars could be viewed more clearly by day or night from the bottom of a well, then it's a trick he probably would not have missed.

EDGAR MITCHELL WAS the lunar module pilot on the Apollo 14 Moon mission, becoming, in February 1971, the sixth human being to walk upon the surface of the Moon. What he experienced on the three-day journey home was, he says, "nothing short of an overwhelming sense of universal *connectedness*." The atoms of his body and the atoms of the spacecraft had the same origins in the furnaces of ancient stars blazing in the heavens all around. He had the deep sense that he and his fellow space travelers were part of a universal scheme, that "the existence of the universe itself was not accidental . . . there was an intelligent process at work." More than that: "I experienced the universe as in some way conscious." He didn't think of it as an otherworldly, mystical experience. "Rather, I thought it curious and exciting that the brain could spontaneously reorganize information to produce such a fantastically strange experience."

Mitchell was raised a Christian. It was a boyhood steeped in mythologies that formed his moral outlook and understanding of the world. As well as the ancient myths of Christianity there were the more recent, and still-potent, myths of the Old West, the stories of frontier, of self-reliance and heroism, good and evil, sin and redemption. Stories of hard lives lived out under the all-seeing gaze of a paternalistic godhead. Myths, ancient and modern, had guided him through childhood to maturity but now, in the spangled depths of interplanetary space, he had a vision of the universe that called for a deeper understanding. "What I needed was a new myth." This new myth, this new story, would be compatible with the cosmology and consciousness of the twentieth century but would also go beyond, reaching out toward a consilience of things cosmic and conscious. "I also realized that myth, when it is new, has always carried the label of Truth." The key, Mitchell thought, was the study of consciousness itself, which, as a field of inquiry, "encompasses all human activity . . . fits precisely into the gulf between the way science looks at the world and the way various cultural traditions do. Mystical traditions assume, implicitly or explicitly, that consciousness is fundamental. Scientific tradition . . . explicitly assumes it is secondary."

The Apollo 11 astronaut Buzz Aldrin was a man steeped in science and engineering. Yet his first act on arrival at the surface of the Moon was to perform a Christian ritual, Holy Communion, in celebration of the supernatural. It was an act of pure *mythos*, of magical thinking, that seems almost in defiance of the powers of *logos* that had delivered him there. The practical, pragmatic, nonreligious Neil Armstrong was having none of it. Ed Mitchell wanted a new myth, a new truth beyond the magic and the science, and the study of consciousness, he thought, was the royal road.

Aliens

THE FRONTAL LOBES OF THE BRAIN, LOCATED JUST BEHIND the forehead, are engaged in planning, problem-solving, abstract reasoning, fluency of thought and speech production. They are often regarded as the seat of intelligence. Neuropsychologists use a variety of measures to assess impairments of frontal lobe function, one of which is the Cognitive Estimations Test. It's a "quick and dirty" bedside test, not particularly sensitive or reliable but, in my experience, it sometimes hit the button with certain patients who were otherwise performing well on cognitive tests. There are different versions of the test but all require the patient to make a range of estimates of weight, size, speed and quantity. Alarm bells ring when someone with an above-average IQ hazards a guess that racehorses gallop at 90 miles per hour, for example, or that there are half a million camels in Holland.

Well before the procedure was formalized in neuropsychology, the great twentieth-century physicist Enrico Fermi was fond of putting his students' frontal lobes through their paces with his own form of cognitive estimations testing. Nicknamed "the Pope" because of his reputed infallibility, Fermi was renowned for his capacity for rapid mental calculation and for cutting straight to the heart of a problem. He was eager to encourage students to develop similar skills. "Fermi Questions," as they are now known, call for rough estimates of quantity that are either difficult or impossible to measure directly but which, via logical analysis, educated guesswork and calculation can be gauged approximately. Some (*How many jelly beans fill a one-liter bottle?*) are less daunting than others (*How many grains of sand are there on all the beaches of the world?*), but all require a similar process of framing and step-wise estimation. One question he put to his students was this: *How many piano tuners are there in Chicago?* Without consulting a directory of Chicago piano tuners, it's unlikely anyone would come up with a precisely correct answer, but by a process of educated guesswork and approximation it should be possible to formulate an estimate of at least the right order of magnitude. There are different ways to approach the problem but one could start with an estimate of the population of Chicago, then take into account the average number of occupants per household and the proportion of households likely to own a piano, factor in the approximate number of orchestras, schools and other institutions with pianos, consider the average frequency with which a piano might be tuned, take into account the number of pianos a tuner would typically tune in a day, the number of working days in a year, and so on.

The same approach can be taken to rather more exotic questions. One day in the summer of 1950 Fermi sat down to lunch with three colleagues at the Los Alamos nuclear research laboratory. One of his companions described a cartoon that had recently appeared in *The New Yorker.* It depicted a fleet of alien flying saucers returning to their home planet. The goggle-eyed aliens are pictured disembarking, each carrying a trashcan marked "DSNY" (Department of Sanitation New York). This would account for the recent spate of flying saucer reports and the strange disappearance of public trashcans from the streets of New York. Fermi joked that it was a good theory. There followed a discussion of flying saucers and the improbability of traveling faster than the speed of light and then the conversation moved on to more mundane topics. But Fermi wasn't done with the aliens. Out of the blue he asked: "Where *is* everybody?"—the extraterrestrials, he meant. He launched into a series of rapid calculations (the details of which are lost to posterity) and concluded that the galaxy should be teeming with extraterrestrials and that planet Earth should have been visited many times over. So, where are they?

I once spent an enjoyable couple of hours in the company of Martin Rees, the Astronomer Royal. He believes there's a good chance of finding evidence for extraterrestrial life—if not intelligence—quite soon. A major advance in astronomy has been the discovery of planets beyond our solar system. It's reasonable to think that some of them will be Earth-like. Detecting intelligence is another matter, but Rees thinks the search is worth pursuing, and he's inclined to believe we are not alone.

So, what about the Fermi paradox? If, as many astronomers believe, the universe is teeming with alien civilizations, how come we see no trace of them? At conservative estimates, there are 200 billion stars in our galaxy, and 200 billion galaxies in the observable universe. To get some imaginative purchase on those numbers I did my own Fermi calculation. Take an average star and scale it down to the size of a pea. Now imagine the Royal Albert Hall stuffed to the rafters with peas. That's a humungous pile of peas, but there are far more stars in the universe than in a pea-packed Albert Hall. Would a couple of Albert Halls suffice? No. Ten? A hundred? Nowhere near. You would, in fact, need something in the region of 400 billion Albert Halls to accommodate all the pea/stars in the observable universe. That's a truly dizzying number of stars, and among the many zillions of planets orbiting those stars there are, surely, a few zillion capable of evolving intelligent, communicative life (and post-biological ways of being). So, seriously, why the eerie silence? And is there not something slightly worrying about it, or at least humbling? Any quasi-omniscient superintelligences out there would seem, at best, to be entirely indifferent to us and our world of woes.

Rees finds the Fermi paradox uncompelling. Absence of evidence is not evidence of absence. So far, the search has been narrowly focused, cosmologically, in the sense that we are pointing our radio telescopes at the heavens with no idea where, or when, to look. It's often described as a needle-in-a-haystack search, but that is to underestimate the task. You'd have to imagine looking for your needle through a drinking straw, and that the haystack is the size of an Olympic stadium. The focus is conceptually narrow, too, because the search for extraterrestrial intelligence (SETI) has an inherently anthropomorphic bias. "We're looking for something very much like us, assuming that they at least have something like the same mathematics and technology." But, as Rees acknowledges, alien intellects could be so different from ours that we might fail even to recognize their signal. "They could be staring us in the face and we don't recognize them." Yes, perhaps there is evidence all around and we just don't see it. We may, as a species, be suffering the cosmic equivalent of Anton's syndrome, the neurological condition in which patients rendered totally blind by damage to the visual cortex believe they can see perfectly well. Perhaps the universe is an act of imagination. Actually, there's no "perhaps" about it. The universe *is* an act of imagination. This is not to say there's no "real world out there," rather that our construction of it is shaped, and inescapably confined, by the powers of the human mind.

Rees asked me what I thought of the eminent physicist Freeman Dyson's suggestion that unusual neurological disorders may give insights into alien minds. Dyson had raised the question in a conversation with his friend Oliver Sacks, and Sacks, apparently, thought there was something in the suggestion. I said I could see what they were getting at, but my experience of working with brain-injured people had led me to a different view. Brain damage and mental disorder open the doors to the haunted mansion of the human mind. We see nothing but the ghosts in our own house— the aliens within, not without.

The SETI pioneer Frank Drake has suggested that the search for alien intelligence is "a search for ourselves, who we are and where we fit in the universe." So it becomes a religion for the secular age, a quest for communion with superior celestial beings through which, perhaps, the supernatural is naturalized. *Silentium dei; silentium* ET. One prominent figure in the SETI community envisages advanced aliens passing on the secrets of immortality. I wondered if Rees would like to live forever? "Yes," he said, "but it won't happen." He added that, when he dies, he would want to be buried, like George Orwell, in an English country churchyard according to the rites of the Church of England. I wondered what implications the discovery of extraterrestrial beings might have for Christianity. If, as Drake postulates, there may be 10,000 advanced civilizations in our Milky Way galaxy, have there been as many Christs, and as many crucifixions? (Hugh MacDiarmid imagines so in his poem, "The Innumerable Christ.") Rees said he knew Jesuit scientists who are quite relaxed about the idea. And then he told me something that surprised me. He said he was a regular churchgoer, albeit one with no specific religious beliefs.

His attitude as a practicing nonbeliever would, he said, be common among Jews. "It's a matter of custom and culture. I grew up in Shropshire in the tribe of the English, part of the Church of England tradition, and I deeply value the aesthetic associations of that culture, the music and the architecture, and therefore have no problem participating in the ritual. If I'd grown up in Iran, I'd be going to the mosque." But he accepted no dogma, and had no time for intellectual theology. What he found very hard to take in some people, who really believe the stuff, was "the craving for certainty." The world is full of mysteries and we should embrace that. He was a little scornful about the physicists' pursuit of the Holy Grail, a *Theory of Everything*. Beyond quantum and cosmos there is complexity, and theoretical physics has little to say about the everyday worlds of biology and society—"an insect is far more complicated than a star or a galaxy, and when you get to the human brain . . ."

Magical Mush

"YOU HAVE WITHIN YOUR HEAD ABOUT A KILO AND A HALF of magical mush. A human brain."

Magical mush? Oh, Jesus!

"It contains a hundred billion neurons."

No it doesn't.

"There are as many neurons in the human brain as stars in our galaxy."

So what?

"The human brain is the most complex object in the known universe."

How do you know?

I'm watching a TV documentary about the brain, and muttering to myself. It makes me aware that Kate isn't here because, by now, she would have told me to shut up.

Next up is the oft-repeated claim that there are more potential connections between neurons in the human brain than there are atoms in the entire universe. Sometimes it's elementary particles, not atoms. This is a claim that may well be true, though I haven't seen a mathematical verification. But there are more potential tunes you can play on a banjo than either of the above. The number of potential tunes is infinite. So what?

Science is now beginning to shed light on our God-given talents, says the voiceover. Surely, if our talents are God-given there's no mystery, is there? God gave them. What has science got to do with it? I'm picking narrational nits here, I know. *For god's sake,* says Kate, sitting invisibly next to me, *give it a rest.*

The screen is filled with ice-blue magnetic resonance images of the brain, lit with red and yellow hotspots. Next a white, plastic-looking brain turns pink as it slowly revolves and the cerebellum starts fizzing with golden sparks.

Stabs of absence.

The Atmosphere

PADDINGTON STATION. I'M SITTING, JET-LAGGED, OUTSIDE the Mad Bishop & Bear, sipping a pint of beer, tapping away at my phone. It's not really "outside." The station roof is overhead. But the roof is translucent and it's a fair simulacrum of "outside." I have half an hour to wait for my train.

A man, fortyish, balding, plump, gray-suited, and bearing two pints of lager, takes the seat opposite. He loosens his tie, exhales profoundly, hopelessly, and gulps one of his pints to the halfway mark. I'm still tapping. I haven't acknowledged him. He pushes the bridge of his glasses up the bridge of his nose, swallows the rest of his drink and slides the second pint toward him. The next time I look up he's halfway down that one.

"Paul," he says, and the sound of my name shocks me.

"I enjoyed your talk," he says, finishing the second pint. "Let me get you a drink." Making a show of checking my watch, I decline. It's now 18:50. My train is the 19:03.

"Sorry. No time." He looks vaguely familiar but I can't place him.

"You've plenty of time," he says.

"I'm getting the seven oh three."

I turn my wrist so we can both see the watch face.

"And it's only six o'clock." He smiles. Now that's very odd. But the station clock confirms it.

He returns with the drinks: two pints each, plus whiskey chasers. I have to laugh. "Hell of a crush at the bar," he says. "Best to get them in." I can see the place is half empty. He winks. He leans across and offers his damp hand. "Mike." The name seems to go with the face but, no, I still can't figure who he is. I've had a sleepless flight from San Francisco and a beer would ease me into a snooze for the last leg of the journey home, but two? Plus a whiskey? I'm not going to sit here getting speed-pissed with a stranger.

"So which talk was that?"

"Goldsmiths College."

I've never been to Goldsmiths. I can't be bothered to correct him, but he's leaving a silence, giving me space to challenge him. Sod that. Your move. I sup my beer. He smiles. I sup. He keeps smiling. I yield.

"To be honest, I don't recall ever giving a talk at Goldsmiths."

"No, you wouldn't."

The whiskey has been downed in one and Mike is on to his second, therefore his fourth, pint. *Goldsmiths?* I tell him I don't follow and he explains that I won't remember the talk I gave at Goldsmiths because I haven't given it yet. He's insane, clearly. Perhaps he's a former patient. The question is, how do I play it? I get on the train and I go home is how I play it. *God, I'm so tired!* I check my watch: 17:55. *What?* I check the station screens: 17:55.

Mike has finished his drinks, so I tell him I'll get another round. "Sure," he says. He rests his fat fingers flat on the table. I go to the bar. I notice the clock behind the bar. Ten-to-six. My thoughts are like planes stacked around Heathrow, waiting to land. I grab the drinks, just the three, just for Mike. I land them on the table in front of him. Nice talking to you, I tell him, but if I get a move on I'll make the 18:03. I give him a matey slap on the back, and I'm off.

The thought planes are coming in to land as I negotiate the escalators and the rush-hour crowds. *Weird.* The time confusion is undoubtedly weird. But there, ahead of me at Platform 5, awaits the 18:03 train to Penzance, solid as a train, the very essence of a train, and if there are no more slippages of time I shall be aboard in two minutes. And, reassuringly, time now seems to be ticking steadily forward.

So what was it? A transient psychosis? Things haven't felt normal. A feeling like something is about to happen or be revealed; something significant; not necessarily bad and not necessarily good. I could see it in the eyes of strangers on the street and I saw it in the psychedelic San Francisco sunset. It started with Marlene and she was at the heart of it, but it was more than Marlene. She saw the signs as well, I think, or some of them. I realize that what I am describing here has a technical, a psychiatric, label. It's a term I've always rather liked from the outside: *delusional atmosphere.*

So who the fuck was Mike? He was a miserable dipso on his way from the miserable office, having a drink or nine to gird his miserable loins for miserable home. He was being sad dipso sociable. He was saying perfectly normal, sensible, drunk things and my agitated, sleep-deprived, paranoid brain was misinterpreting. That's who Mike was.

I'm on the train. I find a seat. I close my eyes. I see Marlene.

MARLENE. SAN FRANCISCO. I entered *the Atmosphere* the night I saw Marlene. With brain and body still on London time, eight hours adrift, I checked into the Orchard Garden Hotel on the edge of Chinatown, had a supper of steam beer and salted nuts from the minibar and slept. But not for long. I was wide awake within an hour and, outside my window, Bush Street was buzzing. I showered, put on clean jeans and a new shirt, and headed across the street for a bowl of pasta at Café de la Presse, and after that I strolled around the block and found myself in Mark Lane and I sensed something was about to happen. I found myself thinking about Marlene. *Mark Lane* must have been the subconscious rhyme-trigger. And then I found myself in the Irish Bank Bar drinking a potent brew recommended by the bartender. And then I found myself looking at Marlene.

There was a sense of inevitability. I didn't tell her I'd just been thinking about her, that I knew something, *something*, would happen, not right then anyway. Later, I did. We did a long, sway-ing hug: friends who hadn't seen each other for a year, who were five thousand miles and more from home. We stepped back and looked at one another. We laughed. We hugged again. She said she didn't know I was at the conference, and I said I wasn't. I had some business down the road in Mountain View. I didn't even know about the conference. It was a big neuroscience meeting and Marlene was presenting her research on testosterone and risk. I feigned interest. She gives pubescent kids money to blow up bal-loons. The more the balloon is inflated, the more money they get, but if the balloon bursts they go home empty-handed. The higher your testosterone levels, the more likely you are to burst your bal-loon. It's quite hard to blow up a balloon to bursting-point. Try it. The tension gets you in the shoulder blades. It's only going to pop. You know that. It's the dynamic between the inevitability and the unpredictability of the shock that gets you in the shoulder blades. Perfect theater. Disappointingly, it turns out that in Marlene's experiment they aren't real balloons, just computer images you pump up by clicking a button.

We talked about her husband, Matt, but not much. Matt is an orthopedic surgeon. I like him. There's a picture of a Ferrari on his office wall. He's forty-five years old but still plays rugby. He drinks me under the table, and he can't blow up a balloon without bursting it. We talked about Kate, but not much. When Kate died, Marlene was going to call; she was going to write; but time went by. I told her I understood and I believe I do. They hardly knew each other. I got us a couple more beers, the one recommended by the bartender, and we talked about other things. I began to think I could listen to Marlene all night.

"Sir." There's a tap on my shoulder and Marlene fades from view as I open my eyes and fumble for my ticket. Settling back, it takes a little while to reconfigure her: the raven hair; the red silk shirt; the sea-green eyes; the dragon eyes; white throat; clavicles sharp as razor shells; the jasmine eyes; the eyes like molten pebbles. *Oh, those eyes!* I can't get her nose right. Or her mouth. If I stop trying to paint her from the watercolor tin of my imagination, if I stop working my brain to the last flecks of pigment and let myself drift with the clatter of the track back to the brink of sleep, then I know she will reappear. And so I do, and so she does, but this time it's not the red silk shirt Marlene, dragon-eyed; this is gray sweater Marlene, unsmiling, on Pier 39 with a Heineken and a fish sandwich which is mostly being fed to the gulls. We haven't spoken for minutes on end. I've agreed with everything she's said and it's killing me. We could have spent the rest of the evening not speaking, and perhaps we did.

I hear an announcement that the next station will be Taunton. I open my eyes and realize I'm sprawled over two seats, using my laptop case as a pillow. I'm paralyzed. My eyes move, but nothing else. I hear everything. Someone in my dream said "Hello," and I'm trying to repeat the word. I try to stop myself but can't. It's no more than a groaning sound at first but when the articulated word finally does emerge it sounds ridiculous. It sounds seedy. Hell-*low*. *Hell-LOOWW*. And I'm staring, head sideways, straight at the woman across the aisle. I can't move and I can't stop myself saying Hello. It degenerates again into a meaningless groaning sound, but by now the woman has moved to a different seat. I still can't move. I—the observing "I"—am aware of the embarrassment, hers and mine. The other me, that thing sprawled across the seats, staring gormlessly sideways, the physical residue of a dream, can't stop groaning. I am simultaneously both selves. Then I wake feeling wonderfully refreshed. A low sun casts long shadows over the rolling fields of Somerset. There was never a more gorgeous, more meaningless sunset.

We pull into Taunton station and a man takes the seat across the aisle where the woman had been sitting. Fortyish, balding, plump and gray-suited, he puts four cans of Stella Artois on the pull-down table. He exhales profoundly, hopelessly. Reflexively, I check the time: 19:47.

"Gather ye rosebuds while ye may," he says. "Old time is still a-flying. You look back, into the past, and the past has gone. The events of yesterday, of last year, of the day you were born, have slipped out of existence. There is nothing."

He offers me a can of lager. I accept it.

"And the future? The same nothingness. Events yet to slip into existence: the events of tomorrow, next year, the day you die."

He leans across the aisle. He latches a fat hand onto my sleeve and swings his fat, sweaty face close to mine. With beery breath he says, "The only reality is NOW!"

"Who are you?"

"Mike," he says, slumping back into his seat and closing his eyes. I have half a mind to make my escape to another part of the train, or jump off and hang around for the next one. We've been stuck here an age with no word as to why, so I might as well be out there strolling the platform, breathing the summer evening air. There'll be another train. No hurry. I have all the time in the world. I look at my watch. Mike, without opening his eyes, says, "Seven forty-seven." He smiles a little smile, which sticks on his face too long, and then he speaks again.

"The present moment is where we are. The only place that's real. For us there is *nothing but* the present moment."

I survey my present moment: the interior of a half-empty First Great Western train. Taunton station through the window. A swig of lager cool in my mouth. People on the platform: a tall man, dressed for a funeral; a young woman shaking back her lush copper hair; a small boy patting a puppy.

"The thing is—and here's the paradox—the present moment is the only reality for us but in the world of physics and cosmology there is no absolute, universal present moment. No moment holds the privilege of unique *nowness*. In physics the *now* is everywhere and everywhen. We inhabit a four-dimensional space-time manifold. All of eternity, alpha to omega, is set in this four-dimensional block. The past, the present and the future are with us always.

That's the physical reality. What's real, the only thing, is the history of the universe as a timeless whole. The passage of time is an illusion."

"You've lost me."

He goes blathering on about the relativity of simultaneity, frozen moments, and timeless quantum cosmology and I'm drifting off again. But this hauls me back . . .

"And the finality of death is also an illusion."

"*What?*"

"Nothing really comes to an end. The flow of time doesn't cease, because there is no flow of time. All moments, all times, are equally real, equally present, including all the moments of your life, which are, from beginning to end, *in place*. Always have been; always will be."

He laughs. And laughs. He becomes hysterical. He laughs till he cries. Then he sighs heavily, hopelessly, and snaps open another can of lager.

The finality of death is an illusion. Can that possibly be true? The moments of one's life imperishably locked for all time in the block universe. Always there. *Always. Finality of death . . . clatter of the track . . . finality of death . . . clatter of the track . . . finality of death . . .*

I've slept. I open my eyes and see that we are pulling into a station. It's Taunton. I check my watch: 19:45. I close my eyes and listen to the passenger announcements and to the sounds of people disembarking and boarding. I open my eyes. Out on the platform there's a tall man, dressed for a funeral, a young woman shaking back her lush, copper hair, a small boy patting a puppy. The train pulls out of the station: 19:48.

The Brain in the Belly

A T 7 A.M., IN THE DARK, LUGGAGE IN HAND, IT WAS LIKE arriving at an international airport. The main concourse was brightly lit with shops and a cafeteria. But this was a hospital. It wasn't the departure lounge we were heading for, it was the operating theater, though general anesthesia is a departure of sorts, I suppose. We were put in a small side room with some other couples to sit in silence and wait. After a while there were drips of conversation:

"Knee?"

"Same here."

Pause.

"Hip?"

"Me, too."

They are older than us. Outside on the ward, behind one of those curtains that screen doctor and patient from view but seem to amplify the consultation, I heard:

"How old are you?"

"Well, now you have me."

My guts were grumbling despite a good breakfast. They seemed to be in a mood of rebellion. Seriously, maybe they were. The enteric nervous system, the "brain in the belly" that regulates the functions of the gastrointestinal tract, is built from the same sorts of nerve cells as the brain up top, a hundred million of them, with the same general functions: input, output and interneuronal crosstalk. It has complex neural circuits, capable of learning and storing knowledge, and a remarkable degree of functional independence.

Sever its links with the mainframe in the skull and it will carry on regardless. So, I thought, why not a primitive mind of its own? *Get on with it*, it was saying, *or get out us out of here.* There's a William Burroughs story, from *Naked Lunch*, about a man who teaches his arse to talk. It starts out as a novelty ventriloquist act but after a while the anus gains the power of autonomous speech and starts ad-libbing and tossing gags around. Before long it's shouting out in the street. It wants equal rights. It talks all day and night. The man can't shut it up. It grows rudimentary teeth and starts to eat. It drinks too much and gets maudlin. We don't need you around here anymore, it tells the man. "I can talk, eat and shit."

People were plucked out of the room at intervals by nurses and anesthetists for their pre-op routines, returning in hospital gowns, faces more strained despite best efforts. The porter arrived to take Kate away at 9:20. He looked at her voluminous luggage. "Blimey! Here for your holidays?"

I was back at my desk by 10:00, trying to work while the surgeon sliced my wife's flesh and hacked at the cancerous bone. Orthopedic surgery is closer to butchery than any other branch of the profession. At 1:00 I called the ward. She should have been out of theater by then, but wasn't, nor by 2:00. The brain in my belly sent pulses of worry up along the vagus nerve, the so-called nerve of compassion, and into my head. But by 3:00 all was well. Kate was back on the ward, way up on the eleventh floor, with a splinted thigh, a new hip, a morphine drip and a view of the sunlit moors. I sat at the bedside reading the sports pages as she drifted in and out of sleep. Then the surgeon turned up to tell us he was pleased with his work. He was cheerful and uplifting, a good doctor who left us tearful with relief. In the evening, back home, I tried again to settle to work but the brain in my belly was up for a beer and I was not going to argue.

That night, the woman who didn't know how old she was got up and looked for her husband. He must be somewhere in the hotel. An amputee screamed with night terrors, and the nurses went about their mundane, extraordinary business. I had my recurrent low-flying airplane dream. I'm in a cavernous airliner with plate glass windows through which I see trees and rooftops perilously close. We're losing power, dropping ever lower. We're going to crash, but not yet. The thing is to try to enjoy the view.

PART THREE

—∞∞∞—

Into the Labyrinth

Oh, Those Snarling Tigers

AEGEUS, THE KING OF ATHENS, WAS CONCERNED ABOUT HIS failure to father an heir so he consulted the Oracle at Delphi and this is what he was told: "Loose not the straining neck of the wineskin, Oh best of men, until you come once more to the city of Athens." Well, figure that one out. Aegeus couldn't, so on his way back to Athens he dropped in on his friend Pittheus, the king of Troezen, who was a very bright spark. Pittheus saw soon enough that "wineskin" was a euphemism for "penis" and that the Oracle was advising Aegeus not to have sex with any women before he got back to his wife in Athens because the next time he "loosed the wineskin" a son would be conceived. But, seeing an opportunity to inject his own royal blood into the veins of a future king of Athens, Pittheus kept this insight to himself. "Sorry, mate," he said, shaking his head. "I haven't a clue. Let me fix you a drink." So he fixed him a drink, and many more until Aegeus was drunk, and then he encouraged him to sleep with his only daughter, Aethra.

Next morning, before going on his way to Athens, and not a little hungover, Aegeus hid a pair of sandals and a sword beneath a rock. He told Aethra that if she bore him a son with the strength to lift the rock she should send him to Athens bearing the sword and sandals to betoken his heritage. What Aegeus didn't know was that during the night as he lay snoring in drunken oblivion, the god Poseidon had also come to Aethra's bed and loosed his wineskin. That's one version of the story, anyway, although it may have been concocted by Pittheus to preserve his daughter's reputation. Either way, in due course, Aethra gave birth to Theseus, who grew into a fine, flame-haired young man, sharp-witted, brave and strong.

One day, when he was about sixteen, a hare he was chasing appeared to take refuge under a large rock, which he lifted to reveal not a fugitive hare but an ivory-hilted sword and a pair of sandals. The time had come, and Aethra sent him on his way to Athens. His route along the treacherous coast road was beset by ruthless robbers and deranged demigods, not to mention a notorious man-eating monster of a sow. I'll give you a couple of examples. There was Sinis the Pinebender, so called because of the sadistic, tree-based methods he had devised for dispatching hapless passersby. The first method, and marginally the more humane, involved bending a pine tree to the ground and placing his victim on the end of it. Sinis would then let go, twanging the poor fellow into oblivion. A variation on this theme required the bending of two pine trees until their tops were close together. The victim's ankles would be strapped to one of the trees and his wrists to the other. You can imagine the bloody tearing-apart as the trees sprang back to the upright. But Theseus was up to the challenge and turned the tables on Sinis, sending him to his death in similar fashion.

Then there was mad Procrustes, who, on the face of it, was quite hospitable and would offer travelers a bed for the night. He had two beds, a long one and a short one. Tall people were given the short one and Procrustes would saw off as much of their legs as required to adjust them to size. Conversely, he offered the long bed to short people and, using a large mallet specially made for the purpose, hammered his guests out until the fit was perfect. From this we get the word *Procrustean*, which often comes in handy in debate because a surprisingly large number of otherwise well-educated people don't know what it means. Both Sinis and Procrustes were sons of Poseidon, so Theseus had slain two of his own crazy half-brothers, assuming, that is, that Poseidon was also his father, or at least half-father (a possible contingency in the surreal world of Greek mythology), and we'll see soon enough that this was indeed the case.

In Athens Theseus was greeted by Aegeus's wife, Medea, the

king himself being away on political business. Medea was a sorceress and had no difficulty recognizing the young traveler, whom she immediately identified as a threat. She knew that, with Theseus around, her own son's claim to the Athenian throne would be scuppered. So Theseus was sent on a mission to restrain the savage Bull of Marathon, the very same beast that had coupled with Pasiphaë to sire the Minotaur. It was surely a hopeless and fatal task, but Theseus relished the challenge and once more proved his mettle. He subdued the bull, wrestling it to the ground by its horns, and then drove the creature back in triumph through the streets of Athens, whereupon he slit its throat in sacrifice to Apollo, the sun god. Aegeus was back in town by now and held a banquet to celebrate the bull's demise. Lacking his wife's powers of sorcery he did not recognize Theseus, and Medea was not done with her murderous scheming. She persuaded Aegeus that Theseus had designs on the throne; it wasn't difficult to fuel the king's fears. He was already in a fairly paranoid frame of mind on account of his brother Pallas (and his fifty sons) being out to usurp him. So he readily agreed when Medea suggested spiking Theseus's drink with a dash of wolfsbane. Theseus graciously raised his cup to salute his homicidal hosts but in so doing revealed a glimpse of the precious ivory hilt of his sword. The truth was out and Aegeus swiftly swiped the poisoned chalice from his lips. Having been exposed as a schemer, Medea fled Athens, never to return, and Theseus was duly proclaimed heir to the throne.

He had done enough by now to establish his heroic credentials but was on the threshold of even greater fame. First off, he helped crush the rebellion of Pallas and his boys. Then there was the business with King Minos of Crete. Minos had a son, Androgeos, an outstanding athlete who had made a name for himself with his victories at the Panathenaia festival in Athens. The Athenians, Aegeus included, had been mightily impressed; perhaps to the extent that the king feared the young man might soon have an eye on his throne. In any event he had set Androgeos the task of pacifying the

Marathonian Bull, whether as a sporting challenge or in certain knowledge that the athlete would meet his end. The bull duly gored and trampled him to death. (There is another, more mundane version of the story in which Androgeos is assailed and murdered by jealous rival athletes. I prefer the royal paranoia, the goring and the trampling.) Either way, his father, Minos, launched a revenge attack on Athens. On top of that he prayed to Zeus for assistance and in response the city was ravaged by plague and famine. Ultimately a deal was struck and Minos claimed an annual tribute of seven young men and seven young women, who were to be shipped to Crete and, upon delivery to the royal palace at Knossos, sent into the labyrinth to meet their doom as fodder for the Minotaur. By the time Theseus arrived on the scene there had already been a couple of consignments, and our hero found himself among the next batch. The youths and maidens were usually chosen by lot but that was not the case with Theseus. In one account he was hand-picked by Minos, so as to inflict suffering upon Aegeus for the loss of Androgeos. Another version has Theseus volunteering, with the heroic intention of slaying the Minotaur and putting an end to the ghastly ritual slaughter.

They set off for Crete under a black sail, in accord with the solemnity of the mission. King Minos was also aboard and before long had started to grope one of the Athenian girls. Theseus defended her and Minos was outraged. What right had a young upstart like Theseus to challenge the great Minos, a son of Zeus? Theseus countered that as a son of Poseidon (the brother of Zeus) he was every bit the king's equal and, moreover, he had the balls to demand proof from Minos that Zeus really was his father. Unfazed, the king called upon Zeus for a sign of affirmation, and the ruler of the gods responded immediately with a thunderbolt. Minos then tossed a ring into the sea and invited Theseus to retrieve it. Without hesitation the young man leaped into the waves as the ship sailed on. His doomed companions despaired for their noble hero, but dolphins appeared from nowhere and bore him down to Poseidon's palace at the bottom of the sea. He returned to the ship, miraculously bone dry, not only with the ring but also wearing a crimson robe and a garland of roses, which were gifts from Amphitrite, Poseidon's beautiful wife.

A couple of days later the ship arrived at Knossos where Minos's daughter, Ariadne, was waiting to greet her father, but it was Theseus, the flame-haired, heroic hunk standing windblown and proud on the prow of the ship, who caught her eye and stole her heart. Under cover of the night she snuck out to declare her love for him through the bars of his cell, and vowed to help him escape. In return Theseus promised that he would take her back to Athens as his wife. He would have said anything under the circumstances, I suppose.

But even if he could overcome the brutal Minotaur—and that was a big "if"—there was still the problem of how to find his way back out of the labyrinth that was the beast's lair. We are talking here about the mother of all mazes, the work of Daedalus, no less. So vast, dark and complex was the labyrinth it would be easier to escape one's own brain. Ariadne went to the architect for advice and the solution that Daedalus came up with was ingeniously simple. I'll give you a clue, he told her, but she was in no mood for games and just wanted the answer. No, this sort of clue, he said, producing a ball of thread.

When the time came for the fourteen young men and women to enter the labyrinth and face their fate Theseus tied the thread to the entrance as he led the way in and unwound the ball as they proceeded. When, deep in the dark maze, they were finally confronted with the hideous hybrid, Theseus valiantly told his comrades to stand back and then leaped into bloody battle with the beast, single-handed. It was a fight to the death from which the Athenian emerged victorious. Finding the way out was now a straightforward matter of following the thread back to the entrance.

Once free they made their way to the harbor and, after smashing holes in the hulls of the other vessels in the Cretan fleet to prevent pursuit, set sail for Athens in the flagship. True to his word, Theseus had Ariadne at his side. She was elated to have won the heart of such a dashing hero. Her neuronal love-chemicals would have been fizzing with the thrill. As for Theseus, what more could a man ask? Still adrenalin-drunk with his conquest of the Minotaur, and with a triumphal return to Athens in prospect, what better than a restorative dose of sun, surf and sex along the way? They stopped for a night on the island of Naxos. Picture the two lovers lying on a beach, naked under the sparkling night sky, distant harbor lights twinkling all colors, revelers' laughter and muffled techno beats rising and falling with the gentle breeze. (Time has no discipline in the sublime.) Flashing, dashing stars hum across the sky from time to time. The first they name Ariadne, the second Theseus, but then they're making love again, and again, and again, and when they are utterly spent they fall into a slumber. Ariadne dreams of being a royal bride and is still dreaming when the sun rises. Theseus is awake by now. He dresses. He covers Ariadne's shoulders with his cloak and he leaves her sleeping. She wakes only in time to see the ship setting sail without her.

Maybe he just forgot about her. He was evidently prone to absentmindedness because, approaching the port of Athens, and with disastrous consequences, he forgot to change the color of the sail. They had set out with a black sail, but were supposed to hoist a white or scarlet one to signal his safe return. Seeing the black sail, Aegeus naturally thought Theseus was dead and, in despair, threw himself off the cliffs into the sea, which from then on bore his name: the Aegean. Another version of the story has him jumping off the Acropolis. In any event, that was the end of Aegeus; Theseus returned to take his place as king, and a very influential ruler he turned out to be. He united the people of Attica and established Athens as the territory's capital, among other things reforming political institutions, formalizing the class system, and minting money. Whether or not there was a flesh-and-bone human being at the Bronze Age core of the story, Theseus became the hero of

Athens, part of the historical lineage of the city and with an influence still strong in the Classical period (500–323 BCE), the political and cultural zenith of ancient Greece, a time and place that shaped us all in one way or another. The Athenians of that era undoubtedly saw Theseus as a historical figure, as we see Julius Caesar and Abraham Lincoln as historical figures. For the Greeks generally at that time the myths were history, or at least legend. Yet they were much more than that. They permeated the culture at every level. But this was also a stage in the evolution of human culture and thought during which the seeds of science were being sown, and science is to myth as oil is to water. Or that, at least, is the way we're taught to think. But I wonder.

We should cast a glance back to lovelorn Ariadne sobbing on the beach as the black-sailed ship goes over the horizon, because her fortunes were about to change. She lifted her head and turned to see a near-naked man in a tiger-drawn chariot. One hand held the four tigers in rein; the other raised a wine chalice, which the charioteer tilted in a beckoning gesture. They gazed at one another but neither spoke. Ariadne looked him up and down. This was Dionysus, god of wine and ecstasy, god of theater, and he had come to take Ariadne as his wife. True, he was not as handsome and rugged as Theseus, but he had a wicked twinkle in his eye, and, *Oh, those snarling tigers!*

The Ship of Theseus

ONE OF THE MOST FAMOUS THOUGHT EXPERIMENTS IN PHI-losophy, the Paradox of the Ship of Theseus, gets right to the heart of the problem of *identity*. In the words of the seventeenth-century theologian and philosopher Joseph Butler, *Every thing is what it is and not another thing*. The question is, what precisely makes something *what it is*; and what makes it the *same thing* with the passage of time? Applied to people, what is it that defines a person and what, if anything, preserves identity over time? What makes you "you," and in what sense do you remain the "same person" over the course of a lifetime?

Here's the Paradox. According to Plutarch, this is what happened to the ship that carried Theseus back to Athens:

> *The ship on which Theseus sailed with the youths and returned in safety, the thirty-oared galley, was preserved by the Athenians down to the time of Demetrius Phalereus. They took away the old timbers from time to time, and put new and sound ones in their places, so that the vessel became a standing illustration for the philosophers in the mooted question of growth, some declaring that it remained the same, others that it was not the same vessel.*

So which side got it right? Those who thought that it was the same ship despite the gradual and, ultimately, complete replacement of the timbers, or those who thought it was a different ship? If just one or a few planks had been replaced, then most people would probably agree that it was the same ship, just as if I have my car fitted with a new exhaust pipe it doesn't even occur to me to think of it as a different car. My old hatchback has in fact undergone a number of replacements over the years: tires, of course; a windscreen; spark plugs; distributor; a door; wing mirror, etc., but still I think of it as the same car. I can't imagine reaching a point where replacing a particular component would make a critical difference and incline me to think of it as another car. So, intuitively, I'm with those who thought the refurbished ship was still the Ship of Theseus. But intuition has its limitations.

The Paradox of the Ship of Theseus was given a further twist by the seventeenth-century philosopher Thomas Hobbes, who imagined someone reassembling the ship using the original materials.

> *If the same man had kept the old planks as they were taken out, and by putting them afterwards together in the same order, and again made a ship of them, this, without doubt, had also been the same numerical ship with that which was at the beginning; and so there would have been two ships numerically the same, which is absurd.*

So here we have two ships. One, the original, gradually being rebuilt with fresh planks; the other built to the same design using the original, rather worn, planks. Which now has the stronger claim to be the authentic Ship of Theseus? Perhaps there is some subtle phasing in and out of existence going on, involving a transfer of the aura of authenticity from one ship to the other. Or perhaps neither can be considered authentic. Either way we are still left with the problem of deciding at what stage in the process the refurbished vessel should be deemed inauthentic. After the replacement of one plank? Ten percent of them? Ninety-five percent? Perhaps it's the idea of "authenticity" that's the problem.

Once Upon a Time

Here's something that happened to me, but I don't think I was around at the time. I was six weeks old. The doctor propped me against a pillow and said to the medical students, "Observe." He fed me a milk preparation from a bottle and stood clear. Within seconds a jet of vomit burst from my lips and hit the far wall at high velocity. "Projectile vomiting," the doctor noted and the students, I imagine, were impressed. I had pyloric stenosis, a blockage of the outlet from the stomach to the small intestine. It required surgery as a matter of urgency, so I was taken to the operating room where a surgical team cut me open to clear the blockage, a procedure performed without the courtesy of anesthesia. This was common surgical practice for babies through to the 1960s, before the development of safe pediatric anesthetics. Did I feel pain? I don't remember. Was there a sufficiently coherent sense of self in that bundle of baby flesh to think and feel: "I am in terrible pain!" even without the words to express it? Young babies are clearly capable of displaying signs of pain and distress, and it takes a sociopathic blockage of empathy to imagine that they are not also *feeling* something at the same time. But who knows? Perhaps there was pain without suffering (if that's at all conceivable). Kate took morphine to quell the pain of her spreading tumors. It didn't so much make the pain go away, she said, as diminish its significance. Imagine being terror-stricken by a vicious, snarling dog, and then realizing there's an unbreakable glass barrier between you. The dog's still there, right in front of you, teeth flashing, but waves of dread reduce to ripples of unease.

So where is Schopenhauer's *moment of great astonishment,*

when consciousness blooms and suddenly we find ourselves exist-
ing after aeons of nonexistence? It's not at the moment of conception. Fertilized eggs don't feel astonishment at anything. Nor
is it the moment of birth. A newborn baby is a bundle of innate
reflexes and behavioral dispositions primed for engagement with
the world, but reflective self-awareness is not part of the start-up
program. Newborns grapple with the blooming, buzzing confusion of a world yet to coalesce through consciousness, and it's a
world that falls properly into place only when we have a sense of
ourselves in relation to it.

I had abdominal surgery without anesthesia at the age of six
weeks and, for all I know, was writhing in agony, but I don't remember a thing. Great pain may have been experienced in that
operating theater but, if so, it was not attached to me as a being
with a sense of its own separate existence in relation to the world.
My Schopenhauerian *moment of great astonishment* had yet to arrive. I'll hazard a guess as to when it did. I'll give myself a creation
myth and say that my conscious self-awareness bloomed with an
illusory bloom at the age of fourteen months.

Airport departure hall. Flight delay. Tired. I've had a pint
to kill the time and now I'm sleepy. A small child sits close by,
fascinated by the floor. I close my eyes and plunge into yellow-gray
mists behind the eyelids, a no-man's-land between exterior and interior. Vague shapes coalesce into a clear, bright scene, a place I've
been before, but not for a while. I can see a door, half open, letting
in light from the street and bringing to life a red flower lying on the
floor. I see my hand reaching to pick it up but the flower eludes me,
or my fingers fail me because, however hard I try, I can't grasp the
stem. It's puzzling. Tantalizing. But the reason I can't grasp the
flower is because it isn't really a flower. It's the image of a flower
woven into the pattern of the carpet. The illusion has fooled me.
How old was I? I open my eyes and watch the kid still doing the

same, grabbing, grabbing at a trompe l'oeil cube in the carpet, to no avail. How old is he? I ask the child's mother. Fourteen months, she says.

So let's say my Schopenhauerian *moment of great astonishment* came at around fourteen months. Whenever I recall the carpet flower I have a sense of seeing, of *being*, for the very first time. Out of nowhere, there I am. A sudden *coming-to*. Some amnesiacs describe a similar experience, a sensation, minute by minute, of emerging out of nothingness into consciousness, *somethingness*, because it's memory that gives us our continuity, that sense of surging through the sea of time, sights fixed on the far horizon, with the days and years roiling in our wake, Helios the sun god driving his golden chariot through the skies above, sea monsters below. We hope the monsters don't rise up from the depths to claim us before the day is out, but we know that night will close in when the golden chariot plunges into the ocean. A child of fourteen months is not yet equipped to make himself the hero of an epic, which is what all of us do in due course. At fourteen months we have only the most rudimentary tools of language and memory. They're too primitive to fashion a grand narrative of the self, but enough, maybe, for a *once upon a time*.

The Guitar

THIS AFTERNOON I DID SOME SORTING OUT OF KATE'S OLD clothes, the ones I've been hanging on to, and took them down to the charity shop. I was breathing them in as I loaded the bin bags. That's it, then. Pretty much everything's gone now, except for the rose-print summer dress and the straw hat, and I can't let go of those. Not yet. Going through a chest of drawers I found Kate's watch, the one I gave her for her birthday the year she was diagnosed. It was still ticking away.

A day of disposal and retrieval. I've dug out the old Gibson acoustic, the one with the chip in the neck from the time Kate grabbed it from me and it fell against the fire grate. So here I am, sad old fart, sat in the kitchen drinking whiskey, middle of the night, strumming my guitar, singing "Hurt," welling up at *Everyone I know goes away in the end*. It hit me in the same spot when Kate was alive, even more so when she was alive, like I was rehearsing my grief. Which is more pathetic? Then or now?

Kate was stoical. She was calm and coping most of the time. Resolute. Death was inevitable and it was coming way too soon, but this was cause for sadness, not anger or despair. There was no deep despair, no bouts of hopeless misery, at any rate, and really not much anger. I recall a flash of *Why me?* exasperation, a bitter, sighing cry out of the blue one day as we sat looking out across St. Ives bay. But that was the only time. In the final months there was some anger directed at me, but not much, no more than you'd get in any marriage at the best of times, I suppose. But close to the end, quarrels are freighted with a pathos they don't carry in ordinary times. Can't you see I'm dying and all you want to do is eat and drink and talk and strum your guitar? *You'll be sitting on my deathbed drinking whiskey and strumming that damn guitar.* I was only trying to do what she was trying to do, I said, which was get on with life. She said, can't we just not eat, drink and talk for once? Can't we just sit and watch a crap, mindless film for a change? So we'd sit and watch a crap film. One of those times—I think it was *Mamma Mia!* we were watching—she said, I'm not very good at this dying business, am I?

Maybe I'm deluding myself but I think there was only once when I really lost it. I'm doing my best, I yelled, I really am. I'm busting a gut. I stopped short of saying *it's killing me,* because, of course, it wasn't. I stepped out of myself and looked at myself, at us. My phantom self floated out and put an arm around me and an arm around Kate and pulled us together, and we found a crap film to watch.

Kate was furious about the Gibson the very day I bought it all those years ago, because I bought it on impulse and it cost a lot of money and we were seriously hard up at the time. I'd gone to collect the cat from the vet's and the vet's was next door to a guitar shop and the Gibson was in the window. I bought it and then went and sat in the waiting room at the vet's with the guitar box resting on my knees and the other people were giving me strange looks like it contained some exotic triangular animal.

Bloody hell, she was furious.

The Ring

IT WAS A SATURDAY MORNING. I PICKED UP MY NEW, WHICH IS to say secondhand, car from the dealer and headed out for the moors under a big, blue September sky. It's trivial things like new cars, new clothes, new furniture, new TV shows that mark the passage of time. I did a new walk. I walked a couple of hours around Grimspound and Hookney Tor, and then I called in for a pint and a sandwich at the Warren House Inn. The log fire was burning, as it always does, even on sweltering, late summer days like this. The "eternal flame" supposedly hasn't been extinguished since the 1840s, when the place was built and they carried the glowing embers in a shovel from the hearth of the derelict old inn across the road and placed them in the hearth of the new one. It's supposed to keep the Devil at bay but must have been burning low the day the landlord blew his brains out in the bar. I took my lunch outside and got stung by a wasp.

On the way home, instead of taking the Plymouth road at Tavistock, I took a right turn at the last second and found myself heading toward Cotehele Quay on the Tamar, near where we used to live, and I found myself walking through the woods to the weir, hardly a soul about, and sitting on the bench we used to sit on, and staring into the green waters of the weir pool for an hour. And then I made my way up the steep path opposite, through the woods to the white gate, and on to the café for a cup of tea. I stopped by at the supermarket on the way home and developed a migraine aura, flashing zigzag scintillations, triggered by reading a shelf sign with parts of the letters missing. Supermarket shelf signs are a hazard for migraineurs. It's a combination of the lettering and the lighting, I think. Back home I took some codeine and napped for a couple of hours to stifle the headache, and then I showered and watched some pointless football, England v. Moldova.

At bedtime I realized Kate's wedding ring was missing from my finger. It might have gone down the shower drain. It might have come off in the supermarket when I was struggling to yank my wallet out of my jeans pocket. It could be anywhere. I've had it on my left hand rather than the right because of the hot weather and it was a bit looser, but not especially so. I searched methodically. Shower drain filter in place, so not there. Kitchen sink likewise. Sofas, back of and under. Pockets, etc. I went out and searched the car. No sign of the ring.

I looked all over for Kate's ring. I drove out again to the moors. I walked around Grimspound and Hookney Tor. I went to the Warren House Inn. No sign, nothing handed in. I left my number, in case. I drove to Cotehele, retraced my steps to the weir and through the woods. I inquired at the café and at the supermarket and they promised to contact me if it ever turned up, and I rang them every other day for the next two weeks to chivvy them. I searched the house forensically, every cupboard, every drawer, every box, twice over. It was gone.

As the weeks went by I thought of getting a replica made. The ring was the work of a Cornish craftsman. Maybe he kept a record of his designs, and so could make another one just the same. But that was a stupid idea. No doubt it would be just as lovely an object, with its tendrils of gold and studs of platinum, but it wouldn't be *the* ring, the original, not the one we bought in St. Ives, the unspoken reaffirmation of vows, not the one I removed from Kate's dead finger. A replica would be just that, a replica, empty of the essence of the original. Things have essences, don't they?

It was not lost on me that I was rehearsing a version of a thought experiment I used to do with my students. It went like this. I think your new wedding ring (it could be any prized item) is exquisite and you kindly let me borrow it so that I can commission an exact copy. I'm delighted with the replica, which is perfect, and call to tell you I'll be returning your ring in the morning. I scrutinize the rings on the palm of my hand and they are indistinguishable, and then I accidentally drop them on the floor. Now I don't know which is which. Next day, I choose one at random and hand it over to you as promised, not mentioning the mix-up. You are happy to have your cherished ring back. It's funny how much you missed it, you say, inspecting it admiringly and putting it back on your finger. It already feels a part of you. My conscience is pricked. I know there's only a fifty-fifty chance that the ring on your finger is the original. So I confess to the confusion. You wouldn't be too pleased about it, would you?

I return to the jeweler. Maybe he can sort this out. The jeweler is meticulous. He inspects the ring closely with a magnifying lens, looking for microscopic identifying marks of his craftsmanship recognizable only to him. He weighs it, having kept a record of the precise weight of both the original and the replica. There was a fractional, humanly imperceptible, difference. This is the original, he confidently pronounces. Problem solved! Unfortunately not, because once outside the jewelers I hold up the ring for the satisfaction of watching it glint authentically in the sunlight but it slips from my fingers and falls under a road roller, which flattens it beyond recognition and redemption. Does it still contain the "essence?" Would you take it over the pristine but "essence-less" copy?

Another forensic search. Every cupboard, box and drawer, twice over. Every pocket. Out again to the Warren House. To Cotehele weir and the woods. To the café. To the supermarket. Nothing. I've lost the ring.

I stand in the backyard under a moonless sky. I take a slug of whiskey.

The Anarchic Hand

IT WAS AS IF THE DOGS WERE GUARDING A DARK SECRET, OR forbidden territory. Get too close and they would tear the flesh from my bones. When I arrived at the house they hurled themselves shockingly at the door, wild, blurred heads gnashing and snarling behind the frosted glass. I was not welcome. On hind legs they are as tall as the man who now appears between them. "Who is it?" That's all he says. He calms the dogs. He lets me in and we go through into the house. No words.

I am sitting opposite Damon in the backroom of his grandparents' house. Damon stubs out his cigarette in a saucer and lights another. He is bare-chested and decorated with tattoos. He hasn't washed for a while. I see it takes an effort of concentration for him to put the cigarette to his lips with the left hand as he reaches for a lighter with the right. He stops, focuses, and works to coordinate the action of lighting the cigarette. Mission accomplished, he exhales and sits in a halo of smoke. I open my case, watched in silence by Damon and the dogs, his barrel-chested Rottweilers, now stationed like sentries on either side of their master. He has a secret they can't allow me to know. If during the course of the interrogation he lets it slip they will leap and rip my face off. That's the thought that crosses my mind. Damon's nan brings us tea and biscuits but I have no appetite. It's the tension, the smell of the dogs, of Damon and worse, of shit, wafting down whenever the old lady opens the door that leads upstairs to her bedridden husband.

The scar across Damon's head is visible through a close crop of red hair. This is where the surgeon dropped through a hole in his skull and found a way down to the corpus callosum, the great bridge of nerve fibers connecting the two sides of the brain. Once there he worked to destroy the bridge so that Damon's left frontal lobe is now disconnected from the right. It's a rare procedure, known as an anterior callosotomy. The idea was to control his epilepsy by confining the spread of abnormal electrical activation to one side of the head. As a last resort, callosotomy can be an effective procedure in some cases when the patient is unresponsive to anti-epileptic medication and when other, less drastic, forms of surgery are not feasible. But not for Damon. The fits are as bad as ever and I am here to catalog the collateral damage. I lay my equipment out on the table and ask him if he minds doing a few tests. "Do we have to?" I promise I'll keep it brief. Through the window I can see sunlight catching the weeds at the top of a black brick wall.

Time drags laboriously in the blue fug of cigarette smoke. Damon is working on an object assembly task, which is a kind of jigsaw puzzle drained of color and interest. I'm testing his visual-constructional skills. He pushes the pieces around without enthusiasm. "This is a fucking waste of time," he says. He is probably right. But then there's a spark of interest. He transfers his cigarette from right hand to left and starts sliding the pieces into place. As the right hand works on the puzzle my attention is drawn to the left. It turns oddly, jerking the cigarette this way and that. It pauses, then, with casual purpose, reaches over and jabs the cigarette into the back of the right hand. "Bastard! Bastard!" Damon is on his feet, gripping the wrist and staring intently at the fingers of his left hand as if eyeing the fangs of a venomous snake. The dogs are barking and snarling. I sit perfectly still, staring straight ahead, avoiding eye contact. My hammering heart could break a rib. Damon releases his grip on the recalcitrant hand. He settles the dogs, sits, and lights another cigarette, which this time he rests on the saucer, and he completes the puzzle. I ask if it happens very often, the left hand acting out of control. "Every now and then," he says. "You have to keep an eye on him." Damon's nan hands me a little packet when I leave. It contains a homemade gingerbread man. I thank her but, later, somewhat guiltily, I throw it away.

DAMON DID NOT intend to stub a cigarette on the back of his right hand, so who did? The action of the left hand was apparently purposeful. It was not an involuntary spasm. The fingers of the left hand manipulated the cigarette and paused (as if for thought) before launching the assault. It was a well-executed maneuver. It seems contradictory to say that the action was purposeful yet unintentional, but Damon, to his mind, was not in control. Some other agent was. The hand, it seemed to him, had a mind of its own, and a shifty, unpredictable one at that: *You have to keep an eye on him.* The episode is an example of *anarchic hand*, which is sometimes observed following surgery to the corpus callosum, although it usually also involves some independent damage to the frontal lobe.

Anarchic hand was observed in some of the so-called split-brain patients first studied by Roger Sperry and Michael Gazzaniga in the 1960s. These were people with medically untreatable epilepsy who had undergone complete surgical separation of the brain's hemispheres, unlike Damon, whose operation divided just the front parts of the brain. (His seizures were caused by scar tissue quite clearly located in the right frontal lobe.) Summarizing their work, Sperry concluded that the cutting of the corpus callosum appeared to cause a doubling of consciousness:

> *Instead of the normally unified single stream of consciousness these patients have two independent streams of conscious awareness, one in each hemisphere, each of which is cut off from and out of contact with the mental experiences of the other. In other words, each hemisphere seems to have its own separate and private sensations; its own perception; its own concepts; its own impulses to act . . .*

He refers to "The presence of two minds in one body, as it were . . ." If that were indeed the case, then the anarchic actions of the left hand might be understood as an expression of the independent will of the conscious, but mute, mind of the right hemisphere.

The "dual consciousness" interpretation has been keenly debated down the years, but the split-brain cases fired the imagination of philosophers such as Thomas Nagel and Derek Parfit, who were interested in consciousness and personal identity. Sperry and Gazzaniga's experiments raised difficult questions about our common assumptions about what it means to be a person and what it means to *have a mind*. Nagel pointed out that, in the split-brain state, "there appear to be things happening simultaneously which cannot fit into a single mind: simultaneous attention to two incompatible tasks, for example, without interaction between the purposes of the left and right hands."

Starting from the commonsense assumption that there are such things as individual minds, Nagel sets out various possible interpretations of the split-brain data. It could be that the patients have just the one "fairly normal" mind, associated with the left hemisphere. In this scenario, the nonverbal right hemisphere is not conscious and its responses are the behavior of an automaton. Alternatively, while the conscious mind is associated with the left hemisphere it may be that the right hemisphere *is* capable of generating isolated conscious phenomena but that these are not integrated into the mind. Then again, perhaps split-brain patients have two minds, only one of which (the left) can talk. Another possibility is that, in normal everyday situations, with the hemispheres working in parallel, they have one mind, which splits in two under experimental conditions, like a river flowing in separate streams around an islet. The two minds then reconvene once the experiment is over.

I won't drag you through the maze of Nagel's exhaustive arguments (his classic paper, "Brain Bisection and the Unity of Consciousness," is available on the internet if you want to check it out), suffice to say, he rejects all of the above options, concluding, "there is no whole number of individual minds that these patients can be said to have." This leaves us with the difficult conclusion that "significant mental activity does not require the existence of a single mental subject," which, concedes Nagel, is extremely puzzling. It's not at all how we usually think about minds. One brain, one mind, one person is the accepted convention. We might conceivably imagine (as in the split-brain cases, or with the drug-induced brain division of a Wada test, a preoperative procedure in which the hemispheres are deactivated in turn for psychological testing) two minds in the same head, but the idea that minds may in some fundamental sense be *uncountable* is a challenge to our ordinary understanding of minds and persons. It may be, says Nagel, that "the ordinary, simple idea of a single person will come to seem quaint someday, when the complexities of the human control system become clearer and we become less certain that there is anything very important that we are one of." He doubts, though, that we will ever be able to abandon the idea of ourselves as singular, unified beings, no matter what the neuroscientists discover and what the philosophers argue. That's how we think of ourselves, and we'll just never be able to get our heads around anything else.

Épater les Bourgeois

THE PHONE RINGS. I REACH OVER TO THE BEDSIDE TABLE and pick up the handset. Nothing. Did the phone ring or did I dream it? I lie awake for a while. The silence deepens and then I hear a stifled sob. A voice, a man's voice, says, forlornly, "I want my mind." So, I'm dreaming.

Am I?

Perhaps this is a false awakening, or a sleep paralysis. No. I can move. It was a hypnopompic hallucination, then. It's five in the morning, just getting light, which is a good time for hallucinations. I think of Larkin's "Aubade," that gloomy dawn rumination on death, *a whole day nearer now* with the curtain-edges growing light.

At eleven I meet Marcia for coffee. We sit in the upstairs room of the coffee shop. There's no one else there so we take the easy chairs by the window, overlooking the street. She suggested meeting at the university but I thought there was a risk of bumping into X, and I was not in the mood for her. Marcia's hair is wet. She arrived after me and got caught in a shower. Now the sun is shining. She's speaking fast, which she has a tendency to do. I'm not listening. I'll listen when I have to. Her mid-Atlantic voice is pitched a fraction lower than its natural register. I cut in. I tell her she's looking great. She's not used to compliments and I know it will catch her off guard and leave her slightly flustered, but pleasantly so. Marcia is unused to receiving compliments about her appearance, not because she is unattractive but because in her world, the world of serious scholarship, such compliments do not often get paid. Actually, what I say is, "Marcia, you're looking great, *by the way*." For reasons I haven't figured out, a casual "by the way" respects personal boundaries while simultaneously breaching them. The compliment penetrates with permission. It's mysterious. Marcia smiles.

"Sooo . . ." This is the signal that we are moving on to discuss her work. She fires up her iPad. "Excuse me," I say, "I need to find the restroom before we get down to business." Why the fuck did I say *restroom*? I'm washing my hands and, taking me by surprise, my reflection in the mirror says out loud, "Go for it." I don't normally talk to myself in the mirror and I'm not at all sure what it is I am telling myself to "go for."

Wagner hated all things French, Marcia tells me, but he was a great hero to the Decadents. She's writing a thesis on music and sexuality at the *fin de siècle*. There was, in the 1880s, a periodical dedicated to spreading Wagner's influence: *La Revue wagnérienne*. Her French accent is impeccable. "They were drawn to him by the urge to . . ."

"*Épater les bourgeois?*" I venture.

"Precisely."

I don't know what *épater les bourgeois* means. *Épater?* Not a clue. But I haven't finished: "*Les faire enrager en louant un ennemi notoire des Français.*"

I have no idea what I'm saying, but Marcia does. "*Exactement!*" she says. "*En particulier, les connotations érotiques et les thèmes illicites dans son oeuvre.*"

I throw up my hands. I tell Marcia I'm sorry but, really, I don't understand a word of French. She's bemused, but less than I am. She laughs and tells me my French is obviously better than I think. "*Peut-être,*" I say, clumsily, and she laughs again.

Marcia wants to *pick my brain*. I like that expression. I wonder who invented it. She wants to know what I make of nineteenth-century ideas about the physiological effects of music. She mentions a writer I've never heard of, Grant Allen. It strikes me as a modern-sounding name, that of a professional footballer rather than a Victorian novelist and science writer. Marcia strokes and taps her iPad and quotes from one of Allen's works, *Physiological Aesthetics*: "Auditory nerves are not likely to be scratched, burned, bruised or attacked by chemical agents, but only to be wearied by over-use or jarred by discordant sounds." She looks up expectantly, but I have nothing to say. I just nod.

Some forms of music were thought to be dangerously erotic, Wagner's especially. It's an idea that goes back to the Greeks. Plato wanted to ban sensual music. Only the Phrygian and Dorian modes would be permitted. The true purpose of music, he argued, was to strengthen willpower and promote masculine characteristics in the interests of society, not to indulge an individual's senses. The nineteenth-century debate, Marcia says, was in some ways just a continuation of the same theme but with a scientific vocabulary. Conservatives, musical and political, considered the music of the so-called *Neudeutsche Schule*—Wagner, Liszt, Berlioz—to be dangerous and subversive. On the other side, and especially among the French Decadents, "nervous music" came to symbolize erotic freedom and the rejection of bourgeois values. Wagner's music above all.

More tapping and stroking and Marcia is reading from *Tower of Ivory*, an Edwardian novel by Gertrude Atherton, someone else I've never heard of. A soprano, Margarethe Styr, has just sung the role of Isolde. Wagner's masterpiece is, she says, "the most licentious opera ever written." Styr "wondered if she were alive or a walking automaton. Her passion had expended itself, the blood had left her brain." She writes to a gentleman friend, Mr. Ordham, "If I cannot make you understand the fearful state of excitement which an opera like Isolde induces, then indeed I hope you will not forgive me, never come near me again. But I fancy you have already forgiven me. I was a wild beast." Ordham ultimately succumbs: "No one was surprised to hear of his illness—brain fever?"

Marcia laughs and she concludes: "So, overstimulation, musically and sexually, could be fatal!" and I, to my surprise, reply,

"*Il y a pire comme façons de mourir.*" "Yes," says Marcia, "I suppose there *are* worse ways to die!" She promises to send me some work in progress, and I promise to read it. I tell her I have some ideas of my own about the physiological effects of music and I can send her some references.

"Sooo . . ." This is the signal that we are winding up our conversation. I'm watching the sunlight on Marcia's still-damp hair. I'm watching her mouth; her eyes. "Sooo . . . Where do we go from here?" My eyes have trekked her contours—I am good at looking without looking—and have come to rest finally, heavily, on hers. And here comes my reply: "*On va s'arrêter là. On a mérité un bon repas. On va se partager une bouteille de vin, ou deux. Après on ira chez moi. Et on baisera tout l'après-midi sur du Wagner.*"

Marcia flushes red. It's instant. She doesn't speak. She doesn't smile. She packs away her iPad. And all the while she doesn't take her eyes off mine. She stands. She walks past me. She goes down the stairs and out onto the street. I'm watching her through the window. My inner voice speaks, except that it doesn't entirely seem to be mine. *Unlike you*, it says, *I paid attention in those boring French lessons.*

Marcia has reached the end of the street. She sits on a bench. Right now I don't seem to have any thoughts of my own, except for the thought that I don't have any thoughts. I'm not even curious as to why Marcia walked out without a word. I'm blank.

Marcia is waiting.

My thoughts regroup. I have said something shocking. I should go and apologize. But for what? "Just go," says my inner voice, and I'm heading out onto the street. "You made a proposition," it explains. "A good lunch, a bottle of wine or two, then back to your place for an afternoon of frenzied fucking." What? *What!* "To the music of Wagner."

Marcia knows I am approaching but she keeps looking straight ahead. I sit beside her but still she doesn't turn her head.

"Marcia, I am so sorry."

The silence goes on and on. She swallows. She swallows again. And then she breaks the silence.

"We're going to do exactly as you say."

But we don't. We skip lunch and there's no Wagner.

Mr. Kafka

I'D BEEN WATCHING THE FRANZ KAFKA LOOK-ALIKE SINCE WE left Paddington. The train was full but the seat next to Mr. Kafka remained unoccupied. The woman had lasted ten minutes, the man about five. He seemed oblivious to both. I could see him mouthing something but his voice was hard to discern against the hubbub of conversation and the ambient clatter of the track. I listened closely. Other solitary passengers sat quietly reading or staring out of the window, but this man was deep in conversation with himself. I assumed he was psychotic and at war with his voices. Perhaps he was, but his tone was more relaxed than you might expect. I'd say he was in good spirits. Even so, he was making people uncomfortable. I registered a hostile glance or two. Why? He didn't seem to be saying anything threatening or offensive. He was only making public what the rest of us keep private and I was intrigued to be dipping into someone else's stream of consciousness. We all have a little inner voice, but to talk aloud to oneself in the presence of others is to breach a surprisingly rigid social code. I wondered if I had the nerve to start talking to myself and realized I didn't.

Plenty of people hold conversations with themselves in private. I'm not much of a self-talker, but that little inner voice plagues me to death when I sit down to write. Faced with blank screen or notepad, the subterranean babbling brook of words bubbles almost, but never quite, to the surface. I can vaguely hear it. But in fact the whole process of writing is akin to talking to oneself. Words pour forth with no interlocutor in sight. The person we are addressing is absent, or our words are intended for no one in particular. And the point of writing, of course, is to take charge of the little voice in someone else's head. Readers happily relinquish control. It feels natural and satisfying to submit to the guidance of another voice and have someone else take charge of one's thoughts. To lose oneself in a book, to be beguiled and steered by the authority of another voice, is almost to enter a state of hallucination.

Inner speech serves the purpose of helping to create and maintain a sense of personal identity; the sense that we are singular and continuous beings. So what happens when the inner voice falls still? Scott Moss, a forty-three-year-old clinical psychologist, was appointed to a senior post at the University of Illinois. On the same day that he passed the physical examination for incoming staff he suffered a debilitating stroke that virtually abolished his capacity for speech. He lost not only the ability to converse with others but also to engage in self-talk. He later wrote of his experiences and the process of recovery, describing the condition of total wordlessness as being like confinement to a continuous present. "I did not have the ability to think about the future," he says, "to worry, to anticipate, or to perceive it . . . I simply existed . . . I could not be concerned about tomorrow." Nor, in his verbally disconnected state, did he have much concern for his wife and children.

I got off the train at Plymouth. Mr. Kafka, I had learned, was going on to Penzance to visit his father, who lived alone, who had been ill, who would be pleased to see his son after so long but not know what to say. Who always kept himself to himself.

I'd Like to Kill Daddy

ONE SUNDAY MORNING, WHEN MY SON NAT WAS FOUR years old, he climbed into bed with his mother. I was downstairs making coffee. "Mum," I heard him saying as I returned, "I'd like to kill Daddy." It was a dispassionate declaration, said serenely, not in the heat of a tantrum or the cool spite of a sulk. He was quite composed. Shouldn't you be repressing this? I thought. Rather than openly contemplating patricide, shouldn't you be *identifying* with your father so as to accommodate your Oedipal impulses? The machinery of his unconscious motivations seemed disconcertingly transparent. In fact, I was less disturbed by what he'd said than by the revelation that there might be a grain of truth in a theory of infant development that I'd always dismissed as absurd and irrelevant. But where else had it come from? I had trained in experimental psychology. I was a professional neuropsychologist conducting research into brain function. There was no place for Freud. And yet, *Out of the mouth of babes* . . .

Freud's ghost haunts us all. Freudian language has seeped into common parlance like that of no other writer since Shakespeare. The core ideas of his psychoanalytic theory have become part of the fabric of our culture. Accounting for human behavior in terms of unconscious thoughts and hidden motivations has become commonplace. We all know about wishful thinking, about denial and defense mechanisms, repression, narcissism, Freudian slips and the anal personality. We all scrape at surface reality for signs of deeper meaning. (I keep typing "Fred" rather than "Freud." Is this a neuromuscular quirk of finger control, or is that my grandfather down there in the basement?) As W. H. Auden wrote, in memory of Freud: *To us he is no more a person / Now but a whole climate of opinion.*

Freud had a clear view of his place in intellectual history. He ranked himself not only with the likes of Copernicus and Darwin but also with that fellow icon of twentieth-century thought Albert Einstein. By his own estimation, he was as much a pioneer of psychology as Einstein was of physics. The Copernican view of the Earth orbiting the Sun, and not vice versa, is irrefutable. Darwin's theory of evolution is acknowledged to be one of the greatest of all scientific achievements. Einstein fused space and time. But what is Freud's standing in this Century of Neuroscience when, through technical wizardry undreamt of in his era, we can view the movements of mind in the machinery of the living brain? What traces of Freud's footsteps do we find in the neural pathways?

In the early part of his career Freud was firmly oriented toward biomedical science. As a student at Vienna University he made meticulous studies of the reproductive system of the eel (sometimes an eel is just an eel) and went on to study neuroanatomy, the field in which he made his first scientific observations. While working in the Vienna General Hospital, mainly with neurological patients, he wrote a landmark paper on the medical and psychological properties of cocaine—substantially based on self-experimentation— and later published a book on neurological disorders of language. As well as being the father of psychoanalysis Freud might also be considered one of the founders of neuropsychology. Inspired by a stay in Paris with the French neurologist Jean-Martin Charcot, who was famous for his work on hysteria, Freud's interests became increasingly psychological and he set out to build a comprehensive theory of the mind. Although believing that, ultimately, such a theory must be grounded in biology, he recognized the methodological limitations of the neurological science that was available to him. Had he been born a hundred years later, with the technical

apparatus of modern brain science at his disposal, he would doubt-
less now be at the helm of an MRI scanner pursuing a career in
cognitive neuroscience.

Neuropsychology has flourished since the time of Freud's
death in 1939. New neuroimaging methods, in combination with
traditional "lesion studies" (which examine the effects of localized
brain damage), have produced ever more refined models of brain
function. We know a great deal more than Freud ever could about
how different neural systems construct the perceptual world from
the raw materials of sensation, and we are mapping the mechanisms that control language, memory and voluntary action. Brain
circuits underlying emotional and motivational states are also
under close scrutiny. The study of emotion, in particular, has been
reinvigorated over the past two decades, so much so that there is
talk now of an "affective revolution" echoing the "cognitive revolution" of the late 1950s and 1960s. Evolutionary theory and experimental neuroscience have combined to produce a framework
for understanding the emotions at every level from the chemical
to the cultural. We know much more, too, about the interrela-
tion of emotion and thought, partly through the development of
cognitive behavioral therapy, which is currently the psychothera-
peutic treatment of choice for depression and anxiety states. Most
neuroscientists are suspicious of Freud, but with its newfound enthusiasm for understanding the emotions and the rapid growth of
research into the brain bases of psychological disorder, it might be
said that neuroscience has nevertheless been moving steadily in
the direction of the central Freudian preoccupations. While neuroscientists and psychoanalysts generally remain, at best, indifferent to each other's concerns, some researchers and clinicians are
working to integrate Freudian theory with brain science. Given
his unfulfilled ambition to construct a biology of the mind, Freud
would have approved.

How does his theory of the mind hold up? His model of the

mind evolved over the course of a long and prolific career, but the idea that our motivations are largely hidden to us, buried in the unconscious, remained a core feature of his theorizing. It is not simply that we find some mental processes difficult to excavate— like time-faded memories—rather that they are actively being denied access to consciousness by a repressive mechanism whose function is to shield the individual from uncivilized thoughts and impulses. According to Freud's later formulations, each of us has three subpersonalities: the id, the ego and the superego. The id is an amoral beast, governed by the *pleasure principle*, driven by instinct and seeking immediate gratification. The ego is the mind's executive apparatus; the rational, decision-making part that enables us to distinguish inner from outer, fantasy from reality. *I want it now*, says the id. Food, sex, whatever, *give it to me now*. *Hang on*, says the ego, *this is hardly the time or place*. But the ego is expedient rather than moral. It would rob a bank if it could get away with it. Moral purpose is the function of the superego. It loads the ego with guilt if it acts out of turn. Without the repressive influence of the ego, acting under the guidance of its moral superior, the forces of the id manifest themselves through fantasy and sexual and aggressive impulses. Mental illness results when repression fails.

Modern neuropsychology provides compelling evidence for unconscious mental processing. For example, it can be shown experimentally that the behavior of brain-injured patients can be influenced by memories that are unavailable to conscious recollection. Antonio Damasio's patient, "David," is unable to recognize photographs of people he has met but, when asked who he might approach if he needed help, will reliably pick out those who have treated him well. This illustrates the now well-established distinction between "explicit" (conscious) and "implicit" (unconscious) memory systems. Similar dissociations are found in our perceptual apparatus such that certain "non-conscious" perceptual pathways have relatively direct access to the brain's emotional memory centers. Moods, memories and emotions can be triggered by events that we simply fail to register at a conscious level. Freud's core contention, though, was that ideas and impulses arising in the unconscious mind are actively repressed and that we carry a whole bag of tricks for the purposes of self-deception: *denial, rationalization, reaction formation, projection*. The eminent neuroscientist Vilayanur Ramachandran, for one, believes that neurology clinics are teeming with examples of behavioral disorder that can be viewed in a Freudian light, so providing evidence of a brain basis for repression and other defense mechanisms. Denial of illness, anosognosia, is certainly commonplace among neurological patients. I recall a conversation with a man, paralyzed from the neck down, who was telling me about his plans to go rock climbing at the weekend.

"Confabulation," the inadvertent construction of false, sometimes fantastical, memories can also be interpreted as a breakdown of the "reality principle" that normally governs the rational ego. It is typically associated with damage to the brain's frontal and limbic systems. Inner and outer become confused; wishful thinking overwhelms realism. A head-injured patient I was interviewing at a rehabilitation center in the north of England casually informed me that he'd been to Australia for the weekend. Didn't I know about the exchange visit? A group of our patients had swapped beds with their counterparts at a hospital in Sydney. He had always wanted to visit Australia, his wife told me. Other flights of fancy included a friendship with John Lennon and an affair with Marilyn Monroe. As well as the confabulations, there were other glimpses of the unrepressed id. "That looks nice," he said, helping himself to a sausage from another patient's plate. "You've got lovely tits," he told the new registrar. This was mild disinhibition compared to the id-like raw aggression and unrepressed sexuality that are sometimes associated with damage to the frontal lobes. Ramachandran believes that observations of neuropsychological disorder not only support Freudian theory but also offer ways of advancing it—"we can carry out experiments that Freudian analysts have only dreamed of."

And yet the dynamic, unconscious mind has a long history in philosophy and literature, traceable from Plato to Dostoevsky, via Shakespeare and Austen. Prior to the development of Freud's ideas, Francis Galton was speculating in the journal *Brain* about "strata of mental operations, sunk wholly below the level of consciousness, which may account for phenomena as cannot otherwise be explained." And, well before Freud, Richard Krafft-Ebing had discerned that unconscious sexual desires were detectable in dreams. That a good part of our mental life and behavior is unconsciously motivated is a long-recognized fact of the human condition. While some of Freud's core themes may well turn out to match the facts of brain function, attempts to transpose the entire gothic edifice of Freudian thought to the laboratories and MRI suites of neuroscience are misguided. It is not just that so much of Freudian psychology seems nebulous and fanciful, it has become clear that in certain regards it is patently wrong. His ideas on female sexuality, for example (penis envy, the inferiority of the clitoral orgasm, and all that), are these days rightly derided.

Neuroscience should resist buying wholesale into Freudian mythology. History may ultimately judge Freud's contribution to science as being a little like that of Franz Joseph Gall, the eighteenth-century phrenologist: wrong in detail but just right enough in general principle (concerning localization of brain function) to build a platform for a more advanced future science. What neuroscience can learn from Freud's grand enterprise, however, is a greater curiosity about the nature of human personality and selfhood. We may know a good deal about specific domains of psychological function: about how the brain processes sensory information, how it organizes language and memory, how it solves problems and guides behavior. But we still have scant knowledge about how such processes give rise to a coherent sense of self, and here I agree with Ramachandran that this is the greatest scientific and philosophical riddle of all. It is quite possible that there is no solution to be found, that the self, like the soul, will turn out to be an illusion. This, after Copernicus, Darwin and Freud, would be a fourth, and possibly fatal, blow to human pride.

The Spaceman's Offer

I GAVE UP LISTENING TO THE WORLD SERVICE, WHICH WAS keeping me awake as often as not, and came almost to enjoy my graveyard-hour appointment with insomnia. It's not so bad if you can settle to watch the parade of thoughts and images as a disinterested spectator, not getting involved, not claiming ownership. Relax, and all sorts of things drift by. You catch snippets of conversation from the day before, or the decade before. Suddenly, you're driving down a motorway, or waiting for a train. *Where am I going?* you wonder. The other night I caught a glimpse of my son Nat and me as if through a time telescope, rolling a giant snowball. The memory was labeled "1986," and if you were to ask me now what I did in 1986, the first thing I would tell you, ahead of anything else, is that I rolled a giant snowball with Nat.

Last night my sleepless brain took me to a conversation I had with Kate. This was years ago, before Mr. Cancer slithered into view. I was saying to her, suppose some benign, super-intelligent being from a distant galaxy shows up one night and says he can grant you another life beyond death. If you accept, there's an important decision to make. When you come to the end of this life you can either arrange for me to join you in the next one, or we part company never to meet again. I got the scenario from a Milan Kundera novel, *Immortality*, which I'd just finished reading. So, there we are, Kate, the spaceman and me. Kate pondering, me wondering. Her counterpart in the Kundera story was incapable of expressing her true inclination. If she chose to leave her husband behind, wouldn't that imply that there had never been any real love between them, that the life they'd had had been built on the illusion of love? So whenever she pictured the spaceman scene she always found herself capitulating and telling the alien yes, of course she would want her husband to be with her in the next life.

I'd put the question to Kate under a starry summer sky, after a sublime meal on the terrace of a French restaurant. *I'd go it alone,* she said. *Wouldn't you?* One lifetime was enough, however much you loved someone.

Vernal Equinox

It's nine minutes past eleven in the morning. From my study, two floors up, I look down across the elongated rectangle of the garden. There is scarcely any sign of movement, but things are happening. A sudden fine drizzle washes the flagstones pink and gives a gloss to the rooftops. The grass needs cutting and now I notice small gold spots where dandelions have sprung. The dilapidated garden table has shed another strip of wood. Just visible, a white rose is coming into bloom. Small signs of disintegration and growth mark the passage of time. The digital clock in the corner of my computer screen tells me it's eleven twenty-one. Time rolls on. Eleven twenty-two. Things are happening. I hear the rumble of unseen traffic. Unseen builders start hammering and shoveling in a neighbor's yard. Seagulls swoop and call. The sun bursts through. Not far away waves are lapping the shore as they have for millennia, day and night, on and on.

Rain falls, grass grows, flowers bloom, builders build, birds swoop, waves lap the shore and the computer clock registers time. Eleven twenty-nine. All the while my heart pumps blood to my brain and my brain surveys the scene. It observes, thinks and records. This is all *stuff happening*, inside my head and outside. There is stuff happening all the time, everywhere. An insect sucks the morning dew; a surgeon puts knife to flesh; an Arctic ice wall collapses; in the depths of space two galaxies collide. All happenings have sub-happenings: the neuromechanical actions that guide the insect's proboscis into the dew drop and the surgeon's knife into his patient's flesh; the thermomechanical changes that lead to the cracking and crashing of cathedral-size chunks of ice into the ocean; the gravitational dance of billions of individual stars and planets swirling in spirals light years wide. Tick-tock goes the cosmic clock and the minutes and the seconds overflow with transformation.

The hard thing, Kate said, was imagining the world going on without her. But it does. The Sun crosses the celestial equator. Spring tilts toward summer. The blackbird sings his heart out at the top of the tree.

The Gypsy

THERE'S A KNOCK AT THE DOOR. IT'S AN OLD WOMAN HAWK-ing stuff. She's been here before. Kate bought some lucky charms, I recall. I'd be entitled to ask for a refund.

"You're a good man," says the old woman, "taking time to talk to a gypsy. Let me show you what I have." She delves into an old-fashioned shopping bag. I could do with some clothes pegs but she hasn't got any. She shows me some bits and pieces of embroidery.

"No, I don't want that."

"How about some lucky charms," she says. "Show me your left hand."

I show my left hand.

"You're a good man," she repeats, "talking to a gypsy."

She places a small shell in my palm.

"That's for good health."

Another one.

"That's for good fortune."

Another.

"That's for the woman in your life."

I tell her I haven't got a woman in my life. She looks me in the eye.

"There was once," she says.

"Yes."

"She was no good, that one. The next one will be the love of your life. That's for her."

No, I tell her, my wife was a very good woman. I loved her dearly.

"The one before her," she says. "*She* was no good."

The peddler backpedals.

"Your wife passed? Well, she's in a good place. She's thinking of you and she wants you to find another good woman. She wants you to be happy."

A fourth shell: "For the children."

I buy the shells.

"Can I tell your fortune, sir?"

"No, I'd rather not know."

Why the left hand? The question occurs to me only after she's gone.

So you're in a good place, then. The gypsy told me. You're thinking of me, she said, and you want me to find another good woman.

Stabs of absence.

Slush Puppie Psychosis

I CHECK THE TIME: 6:13. THE ALARM IS SET FOR SEVEN. FIRST decision of the day: get up or try to sleep some more. I stay put and roll over to find, as quite often lately, a woman's feet resting on my pillow. Fay's feet. She goes through phases of sleeplessness and her solution is to lie with her head at the foot of the bed. There's no logic to it, but it works. She starts off the right way round and I never notice when she switches. Fay's feet are soft and warm. I plant a kiss. She stirs slightly but doesn't wake. Her left foot bears a tattoo below the ankle, TRUST NO ONE. She had it done the day before we met, a year ago today, on Halloween.

I can't sleep. I might as well get up. But then the alarm goes off and wakes me from a dream. That was odd. I thought I'd been awake all the while, but actually I was dreaming. Or was I awake and dreaming at the same time? Is that possible? I reach for my notebook on the bedside table and write: *Awake and dreaming?* I jot down an outline of the dream, which involved Fay being hit by a speeding car. There was a surprising turn of events. The car door opened and I was expecting to see a staggering drunk but out sprang a baggy-trousered clown who ran off down the street honking a horn. It was one of those times when your brain is doing the thinking, the plotting, and it's nothing at all to do with you.

Last night we went to a talk about Buddhist meditation. Fay, bored, whispered in my ear, do men become Buddhists because they can't get laid? I told her to save that one for the Q&A. We didn't stay for the Q&A. Instead we went for a drink at the pub next door. From nowhere, Fay came up with an estimation. She reckoned I had fifteen years or so to live. Not much more. She thought I might make it into my mid-seventies, if I'm lucky. It's because I don't look after myself well enough. Earlier, as we were getting ready to go out, she'd asked did I ever imagine I might just drop dead of a heart attack or a brain hemorrhage. In fact, I often think such thoughts.

Fay has woken up and is now the right way round, facing away, pressing softly back into my body. So what am I to do with those last fifteen years of my life? It's not very long, and the thing about time speeding up as you get older is true, and there's always the heart attack, or the subarachnoid, or the falling under a bus that could, with no warning at all, shrink fifteen years to fifteen days or fifteen minutes. Fifteen years ago was 1999. I see an image of me, Kate and the boys sitting on a beach waiting for the total eclipse of the Sun. Not so long ago, but it looks like another life. How quickly we can become someone else.

It's the half-term holiday and I've agreed to look after my granddaughters, Ebony and Millie. Fay owns a café and sets off for work around eight. I've promised the girls a trip to the park but as soon as they arrive they want to know where Fay is, so first we head over to the café. They want Slush Puppies. I warn them it'll turn their poo blue, which gives them a giggle and an added incentive. I'm not joking, I tell them. Millie sucks fast on her straw but stops suddenly, pressing a fist to the bridge of her nose. The ice-cold drink has given her a headache. Then something strange happens. The pain subsides and she looks calmly, but warily, around the room. "Everybody's talking about me," she says. I ask what she means and she says, "They just are." She's serious. Here's my hypothesis. The straw shoots icy liquid straight to the roof of the mouth triggering a sudden increase in blood flow through the anterior cerebral artery, which is located quite close by on the other side of the palate. The artery supplies oxygenated blood to the medial frontal lobes. It's well understood that sudden disruption of the physiology of the frontal lobes, as happens in frontal lobe epilepsy, can cause transient psychotic experiences such as hallucinations and delusions. I have a name for this new syndrome I've identified: Slush Puppie Psychosis. I write it down in my notebook. Millie doesn't say much for a while. She surveys the scene and sucks her straw, though less furiously. "Are they still talking about you, Millie?" "No, not now."

Fifteen years. The girls would scarcely be out of their teens. I draw them to me and sandwich my head between theirs but they wriggle free. After the café we go to a joke shop to look at the Halloween stuff. There's a spiral staircase at the back leading up to The Chamber of Horrors. A notice warns that it will not be suitable for small children but the shop assistant smiles *up to you*, so up we go. We don't get very far. There's a crack of thunder and a waxwork model of Medusa the Gorgon flashes into view, deathly pale, long white fangs, her head a swirl of latex snakes. The girls flee. They know all about Medusa and her hideous sisters. It's one of my bedtime stories. They know how the handsome hero Perseus was set the task of slaying Medusa, and of how he went to three old crones for advice on how to go about it. The old crones had just one tooth and one eye between them. They shared them around. I make a sucking, squelching sound as I pretend to pluck my eye out. The girls love that. Perseus had various accoutrements: winged sandals to fly him to the edge of the world; a cap that made him invisible; a special sword; and a bronze shield as polished as a mirror. This last one, the mirror shield, enabled him to avoid Medusa's direct gaze, which would have turned him to stone. With cap and shield he cunningly stalked his prey, and with his incredibly sharp sword lopped off her head with a single swipe. Medusa was pregnant by Poseidon, the god of the sea, and their offspring leaped through her gaping neck hole: the winged stallion, Pegasus, and his human brother, Chrysaor. The morning after I first told the story Millie appeared at my bedside and told me never ever to tell it again. It had given her bad dreams. But that night she insisted on it, and the next.

We leave the joke shop and head for the park and the girls

inspect their hands from time to time to see if they're turning to stone. I reassure them that if it were going to happen it would have happened straightaway. I like the park. It's a proper old Victorian municipal park, with a bandstand and tennis courts and well-tended gardens. It's lovely this time of year with the autumnal scents and the dead leaves piling up. We make straight for the children's playground and I find a bench and leave them to it.

Fifteen years. Maybe that's not so bad. The Almighty sends an emissary to make me an offer. Fifteen years guaranteed. I can die peacefully in my sleep on, let's say, 31 October 2029, and will remain in good health until then. Alternatively I can carry on as usual and take my chances. In my reverie the emissary looks like Clement Attlee, the postwar prime minister: small, bald and besuited. Mr. Attlee takes a puff on his pipe, leans forward and says, "Well?" I ask for a minute to mull it over. "Of course," he says. "And, by the way, all memory of this meeting will be erased from your brain." I don't need a minute. No, thanks, I tell him, I'll take my chances. He smiles, we shake hands, and then he floats serenely up and disappears through the ceiling. I feed the date into my phone and find that 31 October 2029 will be a Wednesday.

The first day of November. I wake dry-mouthed and a notch or two hungover. This morning it's Fay's head that rests on the pillow beside me, not her feet. The tumbler on the bedside table is empty. First decision of the day: whether or not to get up and refill the glass. I stay put. I haven't the energy. Nor have I the energy to reach for my notebook to see what it was I wrote in the middle of the night. It was something about the Stoics but I can't remember what, exactly.

I'm not much of a one for celebrating anniversaries. Fay isn't either. We hadn't had any plans for last night, for our first anniversary, and might have stayed home and done nothing but instead we ended up going to the pub where we first met, and later to a restaurant. The Bread and Roses was crammed, just as it was last year, with a lot of people got up in Halloween costume, the usual werewolves and vampires and zombies. We stayed for a couple of drinks and on the way out reenacted our first encounter, only this time I pretended I didn't have the nerve, hadn't had that extra drink to talk to her. This time I just smiled gormlessly and she looked back at me neutrally and I went my way and she went hers, and in that universe, we never got together. Arm in arm, we walked over to the restaurant half a mile away. It was a breezy, balmy evening, more like August than late October. On the way a young drunk called out, "Fuck me, mate, you're punching above your weight! You must be fucking rich!" It was good-natured enough. He said those things because Fay is twenty years younger than me and she's beautiful, and I'm not all that much to look at. But, in any case, she's the one with the money.

Over dinner we talked, among other things, about the frontal lobes of the brain. Slush Puppie Psychosis got us started, but we picked up on a running joke about Fay not having any frontal lobes. I reminded her I'm a neuropsychologist and I know the signs: the impulsiveness; the fiery flashes of temper; the profanities; the disinhibition. She said she's spontaneous, she's passionate, she's uninhibited, which is not the same as disinhibited, and says what she thinks, and was I complaining? We turned over the distinction between *un*inhibited and *dis*inhibited. Case in point: the swim. One warm July evening we were out walking along the banks of the River Dart. Fay decided she wanted to swim, and in the blink of an eye she was naked and wading into the water, copper-colored hair flying loose over brown, freckled shoulders. It was a most natural thing to do and a lovely Pre-Raphaelite sight to behold. But I was fretting about what the passersby might be thinking. There were hardly any, and no one was offended as far as I could tell. Probably it made their day for some of them. So we agreed that she was being uninhibited, joyfully so, rather than disinhibited. I was fretting because my frontal lobes are more conservative than hers, but I was prepared to accept that hers had made the right call. *Just.* If I had joined her, as she wanted me to, that would have been a different matter. For me, a fifty-nine-year-old man with a slight paunch, to be flashing my tackle about in public would have been unseemly and out of character, and therefore disinhibited. An injury to the frontal lobes might make me behave that way, or if I got very drunk, because alcohol recalibrates the frontal lobes. In the normal run of things it comes down to a matter of judgment, which is precisely what the frontal lobes are about. They apply reason and make decisions and regulate behavior. The prefrontal cortex is the pinnacle of human evolution, I said. Those few cubic centimeters of delicate brain stuff are the fountainhead of science and civilization, the dynamo of progress, the . . . But Fay cut through my paean to the god of the prefrontal cortex (Apollo, of course) and pointed out that I was on to my third glass of red, on top of two beers at the pub, and if I carried on drinking like I do, I would be the one without any frontal lobes.

Flesh and Spirit

ONE RAINY SUNDAY EVENING WE FOUND OURSELVES EN-tering the doors of the Brunswick Spiritualist Church just around the corner from where I lived, in time for the start of the 6:30 Divine Service. It was Fay's idea. She was curious. We received a friendly welcome, were handed hymnbooks and took a seat, along with around twenty others in the congregation. There were announcements and notices, a recitation of the Lord's Prayer, a couple of hymns, which I enjoyed singing, and then we got on to the main business of the evening. The medium, a short, plump man in his forties wearing a bulky pullover, beneath which collar and tie, began somewhat unexpectedly with a reading from Osho, the controversial Indian guru. It was something about roses and thorns. To my surprise, he then singled me out from the congregation. There was, he said, someone from the spirit world standing right behind him. Could it be my father's brother? Why doesn't he just ask the spirit visitor, I wonder. Fay looks at me and whispers: *Behave yourself!* I was permitted only "yes" or "no" responses. Suspending disbelief in the afterlife, I had to allow that, yes, it could be an uncle. My father had two brothers, one of them deceased: Uncle Pavel. It occurred to me that Pavel, a Seventh-day Adventist, wouldn't be seen dead in a spiritualist church, so to speak, but I kept that to myself. It wasn't clear why he'd turned up because he didn't seem to have much to communicate. The medium was getting signals of a bad back. Did he have a bad back? Well, I couldn't say for sure. Possibly. The medium had feelings of discomfort in the throat. Any problems there? Not to my knowledge. Or, said the medium, perhaps it was me. Perhaps it meant I was blocking something. The cramped throat was symbolic. Was there something I was finding it difficult to communicate? No, I don't think so. We weren't getting very far. Now the medium was getting a visual image. He's slightly built? Yes. His mouth and mine have a similar shape? No. His jacket sleeves are too short for his arms. Was that characteristic? No, but I do seem to remember he wore his trousers at half-mast. Finally, I am told that my uncle really wasn't as dour as I thought he was. He wants me to know he had a fun side. If so, I'm thinking, you kept it well hidden, for the most part.

I DON'T BELIEVE there is an afterlife. But there was one night, about a year after Kate died, when I experienced a flicker of doubt. It was altogether an absurd night. I'd gone to bed around midnight, soon fell asleep but then woke at hourly intervals. I heard shouts, which at first I thought were part of a dream, but no, they were real enough. A man was bellowing, *No, no! Oh, no, no, no!* There was no fear in his voice. It was a lament and I pictured him on the nearby hill that overlooks the city and the sea, fists clenched, looking skyward into the rain. I gave him a story. I gave him several stories: 1. His girlfriend had dumped him; 2. He'd come home in the early hours after a drunken row with his wife to find her swinging from a rope; 3 . . . Well, they went on and on.

I slept again, I woke again. Two-thirty. I picked up my phone and saw there was a new email. It was from a woman I scarcely knew. She came to a talk I gave and buttonholed me for half an hour afterwards with intense and urgent questions about neuropsychological assessment. She said she would send me a paper on psychometrics, which she did and which I knew I would never read. The message accompanying the paper was almost as long, and as deadly academic, as the article itself. This new message was considerably shorter: *Paul!* That was it.

I had started dreaming more often of Kate. She had taken a

while to reappear after the first, strange dream a couple of days after she died. I had tucked her into bed and brought her a glass of water. But, I suddenly realized, "You don't need water now you're dead." "No, you're right," she said. And then she was hovering over me, bare-breasted and beautiful, consumed by flames that didn't seem to burn her. She was smiling. She was gone. It was months before she returned and when she did she was her old self and it seemed perfectly natural for her to be around. They were deathless dreams. Now here she was organizing a trip to Scotland with the boys, who were small children again. Had I agreed to this? I didn't want to go. I had a new cherry-red Fender bass guitar and preferred to stay home and play that.

I woke again with the ping of my phone taking delivery of another email. Hadn't I turned it off? This message was from Kate and, no, I wasn't dreaming.

We had endless conversations in the last weeks of her life, often more lighthearted than you might imagine. One of them went something like this:

"You'll find someone else soon enough when I'm gone."

"I'm in no hurry."

"You'll need a companion."

"No I won't."

"A lot of widowers find a new partner within a year or two."

"And I'm not going to think of myself as a widower, that's for sure. *Widower* for Christ's sake!"

"How about . . ."

And then she reeled off a list of potential partners who she thought might suit me and, more to the point, of whom she approved. I told her I was going to play the field.

"At your age!" she scoffed.

"Just watch me."

"Yeah, from the afterlife!"

"There isn't one," I reminded her.

"But what if there is?"

"I wouldn't want you spying on me for the rest of my life."

"But just suppose there is. Should I send you a sign?"

"OK. Why not? Just the one, though." One irrefutable sign.

"Just the one, then."

I looked at the newly arrived email from Kate Broks. I opened it. *Amazing!* it said. *You will feel something new and unforgettable.* But this wasn't the "one irrefutable sign" we'd joked about. It was Viagra spam. One of her email accounts, the one I hadn't been able to close, had been hijacked and harvested. Everyone on the contacts list was going to get this shit. And they did.

IT'S SATURDAY NIGHT and Fay and I are driving home from a choral concert at Buckfast Abbey. I've just told her the email story.

"But maybe it *was* her," she says.

"Oh, come on!"

"Maybe that was the only way she could do it."

"Through Viagra spam?"

"She was having a laugh. And maybe that little message says it all: It's amazing. You feel something new and unforgettable."

The rain is relentless and I am thinking, through the relentless rain and the headlights and the taillights: *What if it really was a message from beyond?* How wonderful to experience a fundamental shift in one's view of the world; to see that there might, indeed, be a life after death. *Something amazing. Something new and unforgettable.* What would it take? It would take more than spooky Viagra spam.

Later Fay reads from a book picked from the pile at the bedside. It's about paranormal experiences and claims for evidence of an afterlife, stuff I've been dipping into since the visit to the spiritualist church. I'm listening as much to the soothing quality of her voice as to the meaning of the words, the talk of spirits and presence, apparitions and physical manifestations. The bedside lamp suddenly dims. We look at one another and laugh. It's a physical manifestation, I suggest. Then it dims down a level further. It's touch sensitive. If you tap the base of the lamp the bulb dims or brightens in a stepwise fashion, but now it seems to be doing so of its own accord. It dims a further two steps down to darkness and then climbs back up again to full brightness. We are now looking at one another but not laughing. Hard-nosed rationalist though I profess to be, I am spooked. We both are. I unplug the lamp. If there's a fault I don't want it flashing on and off all night, and if it switches itself on without the benefit of electricity, then it really would be hats off to the supernatural. It doesn't, though I lie awake for a while half believing it might.

I wake in the early hours and, in the semidarkness, I can see a woman on my right-hand side propped against the pillow. Her hair has fallen across her face and she's looking slightly away and down, like she's reading a book or checking a mobile phone. It's too dark for reading and the faint green glow on her face suggests the reflected light of a phone screen. First thought: Why is Fay checking her mobile at this hour of the night? This is followed by the sudden realization that Fay is behind me on the other side of the bed. At this the apparition vanishes. The green glow, I can now see, is from the digital display of my bedside radio. The hair falling across the face is the shadow of a T-shirt flung across the bedhead. How real she seemed.

Not Quite the Right Universe

WHY ME? WHY MY PARK BENCH? THERE WERE PLENTY OF free benches. The fat drunk plonked himself down and pulled a can of Special Brew from a supermarket bag. He'd had a few already, I'd say. He offered me one. I declined. He seemed vaguely familiar, but I couldn't place him. He gave me a *don't you recognize me* sort of look, then put his hand out, and I reluctantly shook it. "Mike." "*Mike!*" He winked. "Just testing." I was about to go on my way but he laid a hand on my arm. "Isaac Newton," he went on, "Isaac Newton was a genius but he died a virgin. He was a sad fucker." I was taken aback because I'd just then been thinking about celestial mechanics, not about Newton exactly, but about the rotation of the Earth, and the Moon's gravitational forces turning the waters of the estuary, and the Earth orbiting the Sun and how, when you think about it, it all seems so precarious, I mean, being held in place by this invisible force called gravity. What would it take for the Earth to shoot off at a tangent into the deathly cold depths of space?

"Newton," he tells me, "was a horrible man. Spiteful, twisted and cruel. He wrote more about the Bible and alchemy than physics. He'd have been better off chasing women. What a waste of a life." He took a long pull on his Special Brew. Hardly a wasted life, I suggested, if you've come up with some of the greatest theories in the history of science, and he said someone else would have figured out the laws of motion sooner or later. "I'll give you a choice," he said. "In your next life you can either be a spiteful, twisted, cruel and unloved Bible-obsessed virgin who comes up with a new theory, or you can have lots of sex, love and all-round emotional fulfillment." He had a point. "Now, Einstein," he said, reaching into his bag for another lager. "Different altogether. I don't mean just as a person—I mean as a thinker. If Einstein hadn't come up with the general theory of relativity . . ." He stops in his tracks, looks past me, and does an odd double take. He gets up, plastic bag in one hand, lager in the other. "This bench . . ." He's off now. "Sorry, mate," he calls over his shoulder. "Not quite the right universe."

IF RELIGION LAYS claim to certainty and truth, then one view of science is that it does quite the opposite. I hadn't thought of it that way when I was working as a scientist, but science is a sort of craving for *uncertainty*. It marches on, as the physicist Max Planck said, "funeral by funeral," old certitudes slain by successive generations of theory and experiment. On it goes with no final destination in mind and no inkling of the wonders and mysteries that lie ahead.

Even for physics, the king of the sciences, knowledge is always provisional and it is rash to claim an ultimate theory of anything, let alone everything. At the cusp of the twentieth century William Thomson, Lord Kelvin, a preeminent physicist and engineer, proclaimed that "the more important fundamental laws and facts of physical science have all been discovered, and these are now so firmly established that the possibility of their ever being supplanted in consequence of new discoveries is exceedingly remote." All that remained was to refine measurement to ever greater degrees of

accuracy. "Our future discoveries must be looked for in the sixth place of decimals." Kelvin also thought X-rays were a hoax, until he saw the evidence with his own eyes, and he scoffed at the possibility of aircraft ever becoming a practical mode of transport. From the perspective of classical physics, Kelvin's view of the regal procession of physical science seemed clear enough, but there were mysteries and wonders beyond imagination just around the corner. The apparently fixed frames of physical science were about to be reconfigured by Einstein's Special Theory of Relativity, and then completely overturned by a revolutionary new way of conceiving the physical world: quantum mechanics.

The contradictions of classical and quantum physics are comparable to some deeply conflicting, indeed irreconcilable, views of selfhood. One might say there is a "classical" view of what it means to be a person, and there is a "quantum" view. The classical self would correspond to the intuitive, commonsense idea of there being an essential core of identity that gives us coherence and continuity down the years (a "soul" if you will), whereas with the quantum self there is no inner essence to claim control and ownership of actions and experiences. Human selfhood consists in nothing more than a long series of interconnected mental states with nothing at the core. (We earlier considered these opposing views in terms of Ego Theory versus Bundle Theory—page 118). David Bohm in the early 1950s and Danah Zohar forty years later each found seductive resemblances between the "new physics" and the psychology of self-awareness, but they went beyond analogy. They sought to explain consciousness in terms of actual quantum mechanical features of brain activity. The eminent mathematical physicist Roger Penrose has also speculated that quantum effects may play a role in creating the unity of conscious experience. Although I don't go along with any of this, I think our conceptions of selfhood and our understanding of the physical world are riven with peculiarly similar ambiguities and that in each case we are faced with reconciling the apparently irreconcilable.

In the classical, Newtonian scheme of things, material objects have a determinate, individual existence. They have definable locations and follow definable paths at specifiable velocities within fixed and absolute frameworks of space and time. This fits with our everyday intuitions and experience of the world. Whereas the classical world is deterministic, even when reflected in the fairground distorting mirrors of relativity, the quantum world is probabilistic. Nothing is certain, which is to say that, *in principle*, nothing is certain. It's not that we lack the technical and conceptual tools for a complete description of objective, physical reality, just as we lacked the ability to observe the moons of Jupiter before the invention of the telescope. Rather, at the deepest levels of description, there is no fixed, objective reality. The world around us melts away to a realm of ghosts. At the atomic and subatomic levels, things don't hang around like pebbles or planets waiting to be discovered and described. They hover between existence and nonexistence, achieving actuality and presence only through the process of observation itself. Even then there are limits to the scope of description that defy ordinary intuition. Physical phenomena manifest themselves in fundamentally contradictory ways. For example, light can accurately be described as both a stream of tiny particles (photons) like peas from a peashooter, and as a wave, which is to say a pattern of energy passing through a medium (air, water, etc.), like ripples on a lake, or whipped undulations on a length of rope. There is nothing mysterious about individual, localized objects (which is what particles are) behaving in a wave-like fashion. Think of a "Mexican wave" in a packed sports stadium. Here the "medium" is the crowd and the wave is a succession of individual people (the "particles") jumping up to fling their arms in the air and then sitting back down again. Each particle/person remains located at the same seat and no one is carried around the stadium. There are particles and there is a wave. At the quantum level, however, individual particles have been shown to have intrinsic, wave-like properties, with both a distributed and a localized presence.

This is tantamount to the Mexican-wave participant staying put in the same seat and, *at one and the same time*, being whisked around the stadium. Or try to imagine walking through two different doors at the same time. You are in good company if you find this inconceivable, but that's the way it is in the quantum world.

In a famous series of lectures on the character of physical law delivered at Cornell University in 1964, the great physicist Richard Feynman put it this way:

> *I think it is safe to say that no one understands quantum mechanics. Do not keep saying to yourself, if you can possibly avoid it, "But how can it be like that?" because you will go "down the drain" into a blind alley from which nobody has yet escaped. Nobody knows how it can be like that.*

This chimes, as we have seen, with J. B. S. Haldane's famous assertion that "the Universe is not only queerer than we suppose but queerer than we *can* suppose."

Up until the revolutionary physics of the early decades of the twentieth century, the facts of science were broadly in accord with our ordinary intuitions about the material world, ordinary intuitions deep-rooted in our evolutionary history. Our understanding of certain basic features of the everyday physical world of surfaces, objects, places, motion and events is hardwired into the brain. Painstaking psychological experiments using indirect physiological and behavioral measures of attention and response have confirmed, for example, that small infants know intuitively that toys knocked off a table invariably fall to the floor rather than flying up to stick to the ceiling. They know that solid objects do not pass through each other or suddenly disappear into thin air. In short, babies show puzzlement if objects and events appear to defy the laws of nature. We also understand from an early age that time flows forward and is irreversible. Yesterday is behind us; tomorrow lies ahead. While anticipating the arrival of tomorrow we understand that yesterday is irretrievable. Likewise, we know that events may be separated in time *or* simultaneous; one or the other, not both. I take a sip of tea. The action coincides precisely with the movement of the second hand of my watch, which coincides with the changing of a traffic light in the city center, and the chiming of Big Ben, and the breaking of an ocean wave on a Madagascan beach six thousand miles around the globe, and a small meteor burning up in the Martian atmosphere, and a glint of moonlight on the waters of a world at the edge of the galaxy, and so on. Within the cosmic Newtonian frame of absolute space and time, and the drift of ordinary human intuition, these happenings are simultaneous whatever the vantage point of the observer. But that is not the case. Einstein showed that the timing of events is relative to the observer. A satellite in geostationary orbit midway between England and Madagascar would register Event A (the chiming of Big Ben) and Event B (the breaking of the ocean wave) as being simultaneous, but for an observer traveling at close to the speed of light along the London–Madagascar axis the events would be separated in time; A would happen before B (or vice versa depending on the direction of travel). None of these views—simultaneous; A before B; B before A—is any "truer" than the others.

Trying to get a purchase on the core concepts of relativity I inhabit Einstein's thought experiments. I travel at inconceivable speed in pursuit of a beam of light; I observe, in the slow motion of imagination, beams of light traversing lengthwise the carriage of a train, first in one direction, then another; I watch a young astronaut bidding farewell to his twin brother as he prepares to set off on a close-to-light-speed round trip through interstellar space and returning, still young, to find his twin is now middle-aged. It takes me beyond the ordinary intuition that enables me to picture planets in orbit around the sun, obedient to Newton's laws of motion and gravity, but I can, just about, push my imagination into relativistic realms. Quantum physics is another matter. Einstein, whose work laid the foundations of quantum theory, was always uneasy with its bizarre, counterintuitive implications. He hit upon a metaphor of madness to express his discomfort. "This theory," he said, "reminds me a little of the system of delusions of an exceedingly intelligent paranoic, concocted of incoherent elements of thoughts."

There's a tantalizing pleasure to be had in the unfathomability of quantum physics. But what if we ourselves are unfathomable? Not in the sense that so-and-so is "hard to fathom" or "a bit of an enigma," but rather that human beings are, to the human mind, fundamentally, intrinsically incomprehensible. We might get glimpses of what we fundamentally are, as through thought experiments we get glimpses of the quantum world, but only glimpses.

WALKING THROUGH THE park this afternoon it was trees and grass I saw, and benches; a bandstand; flower beds. There were children on their way home from school; dogs chasing balls; well-wrapped old people from the residential home in a convoy of wheelchairs; war memorials; tennis courts; dockyards in the distance; a bright stretch of estuarine water with woods and hills beyond; and all this under quick clouds and a blue sky. Everything was familiar and settled. It was an autumn afternoon in an English municipal park. Devonport Park. There was no relativity or quantum strangeness here. Yet, making my progress through the scene, I was aware of something stranger still at the very core of all that seemed familiar and settled. There was a human point of view: mine. No doubt there were other points of view absorbing the sights and sounds of the park, those of the children and the dog walkers, for example, even of the dogs, but I could access only them through inference. The trees, the sky, the birds and the squirrels were directly accessible, however. They were dancing on my retinas, and flowing through the visual systems of my brain. The neural workshops inside my head were making them up as I went along.

I SIT ON the bench where Kate and I used to sit in the last months looking out over the estuary.

Kate's voice:

"You know we're close to the end now, don't you?"

"Yes."

"A few months. You understand?"

"Yes."

"Sometimes I think it hasn't sunk in. You're not denying it exactly but . . ."

"No, I do understand."

That was the gist of one conversation and this bench is the gist of the bench where we sat, but only the gist. It seems to have been moved twenty yards to the left of its original position and, of course, I can't be sure if it's the very same bench. They're all much the same. Anyway, I walk across to where I thought the bench used to be positioned. Each one is set in a recess off the main path, but there's no recess here. So maybe my memory is doing mischief. I return to the bench that maybe was, but maybe wasn't, the one we used to sit on. The uncertainty is unsettling. I want to *know*. A Tourettic, white-haired old woman being pushed along in a wheelchair goes by, robotically chanting "Fuck off, fuck off, fuck off." I get a smile from her carer, a girl in her teens who seems to be taking a vicarious delight in scattering the obscenity across the peace of the park.

This world of trees and rivers, clouds and sky is an ordered world, a clockwork world. The rivers flow into the estuary, the waters of the estuary merge with the sea. Time passes at a steady rate in one direction. The tide, now low but rising, will at a specified moment turn again, today a little after 9:27 p.m. I've just checked on my smartphone. Sea level will rise and fall 4:60 meters as a result of the gravitational and centrifugal forces at play between the Earth, the Sun and the Moon. The Sun rose at 7:50 a.m. and will set at 6:12 p.m. The waxing, crescent Moon rose invisibly at 11:39 a.m. and will set at 8:18 p.m. It's all beautifully, predictably Newtonian; captured and contained within the laws of motion and gravity, that mysterious contract between the abstract and the concrete through which numbers do not merely find correspondence with the physical world but reveal and *prefigure* it. The Sun will rise at 7:52 tomorrow because the calculations tell us so, and the two-minute difference between sunrise today and sunrise tomorrow is due to a precisely, numerically understood adjustment in the tilt of the Earth as it spins in orbit around the Sun.

So, I watch the world go by. I watch long enough to see the shadows grow as the sun drops and I picture the Moon in the wings waiting to appear. I contemplate the grid of gravitational forces working to turn the waters of the estuary, and the unperceived backward rush of the bench I'm sitting on: over 1,000 kilometers per hour as the Earth rotates eastward. It's all Newtonian. So, too, are events on the human scale, and the squirrel, bird, tree and blade-of-grass scale. The sun will rise tomorrow, for sure, at 7:52. Its elevation at noon is already scheduled. A ball launched in a certain direction at a certain angle and velocity for a dog to chase follows a predictable trajectory and lands at a predictable spot. The Newtonian world is, in principle, utterly predictable. The past is a cascade of events frozen in lost time and the future is determined. Effects follow causes. This is the world that the human mind has evolved to comprehend and negotiate. But come down from the interplanetary level; descend through the human scale; go lower, deeper, smaller into the tissues and cells of the skin or the fibrous structures of a blade of grass and deeper still to the level of molecules and atoms and you enter a world that does not follow the Newtonian rule book. This is the quantum world, a wonderland where nothing is certain, where, in contrast to the Newtonian world, everything is, in principle, *uncertain*.

IT'S BUGGING ME now. Is this the same bench, in the same place? Either it is or it isn't.

Kate's voice:

"Are you sure you understand?"

"Yes, I do."

"Perhaps you need to talk to someone."

"Why?"

I don't so much care if the bench has been moved, or even if it's a different bench in a different location. But I want to know one way or the other. I want to locate that conversation. Although unlikely in the extreme, it's within the bounds of quantum possibility that the old bench dematerialized soon after my wife and I made our last visit and then rematerialized in this new position. It would have confused the park managers at first, but they would have put it down to a misunderstanding with the contractors and then thought no more of it. It's also within the same remote bounds of quantum possibility that Kate could materialize beside me right now and we could try to figure it out together. I hope she doesn't. It would confirm for me that I was mad, even if I'm not.

Kate's voice:

"I thought it was more in that direction."

"So did I."

"Looking over to Kit Hill."

"Yeah."

"How are you coping?"

"Fine."

"So you don't know for sure."

"What?"

"If this is the same bench."

THE HUMAN BODY is undeniably and inescapably an object in the Newtonian world and subject to the laws of motion and gravity like any other material object. If I climbed a tree, hung from a high branch and let go, I would fall in the direction of the center of the earth, accelerating at the rate of 9.8 meters per second squared, like a lump of iron. But there's a difference between me and an inanimate object like a lump of iron. I am animate. I move of my own accord. And there's a difference between me and animate objects like moths and earthworms, which also move of their own accord but whose behavior is coded and constrained entirely by their genes. The moth fluttering fatally into a candle flame does not stop to reflect and consider other courses of action. It is executing a genetically programmed and unalterable pattern of behavior. We human beings are not only animate, we are reflective and deliberative. We move of our own accord (like moths but unlike billiard balls), but we also have the freedom to choose and alter our patterns of behavior. We have free will.

So, I can choose to stay seated on this bench or I can choose to stand. I can choose to walk along the path to the left or to the right and go anywhere I want. It's true that while hurtling toward the ground in free fall from the top of a tree I wouldn't have any say in the matter. I couldn't change direction or defy gravity by slowing my rate of acceleration. But I do have the freedom to stay seated on the bench or to get up and go, with no external forces pushing or pulling me like a billiard ball and unfettered by the overbearing influence of genes and instincts. Does this mean that in some ways I am locked into the laws of nature (falling from a tree) and in other ways not (exercising free will)? I can't defy gravity, but unlike a leaf floating on the breeze I'm free, within a limited (but incalculably wide) range of options, to choose the direction I take. That's how it seems. I could sit on this bench for the rest of the day or I could get up and go somewhere, do something. There are no obvious physical determinants within my limited but incalculably wide range of behavioral options, although there are social and ethical constraints. Most of the things I would be likely to do would be mundane and morally neutral, like walking home and having a cup of tea, or catching a bus into town. But I could, if I so wished, go over to the charity shop and give them my new jacket and all the money in my wallet. That would be a "good" thing to do. Or I could do something "bad," like sneering contemptuously at the next person who smiles at me and barging them out of my way. I could do a bit of shoplifting in the co-op on the way home, or I could do something just plain ridiculous. I picture myself stripping off and strolling around the park stark naked singing "The Banana Boat Song." *Hey mister tally man, tally me banana.* I know I won't do such things, because I consider myself a "decent" person, willingly bound by social conventions and the laws of the land as well as by the laws of nature. But I *could* do them. It's commonplace for people to flout social conventions and break the law—and I have done so—but no one has ever contravened the laws of motion and gravity, nor ever will.

But there's a problem here. In the clockwork Newtonian universe every effect has a cause. One thing necessarily follows another. There is no scope for *acausal* happenings, that is, for events, *effects*, without causes. A long chain of causes and effects has led to my sitting on this bench this afternoon. These include the mechanical movement of my muscles and joints in the process of walking, and the expenditure of energy that entails.

I try an exercise. I decide that I shall rise from the bench either at precisely four o'clock or at one minute past, but I won't take the decision until 3:59 and 59 seconds. I don't have a preference. It is now 3:55, more or less. I really don't know which it will be, 4:00 or 4:01. I watch things and think things and I pay attention to what I am seeing and thinking. I admire the dexterity of a squirrel manipulating an acorn. I check my watch: 3:58:45. I recall a conversation I had with a friend yesterday. What exactly did she mean by *I have heard your words?* That was an odd formulation. *I have heard your words* (and they are hollow . . . insincere . . . self-delusional); *I have heard your words* (and they are honest . . . touching . . . heartfelt). I can see her face in my mind's eye and there's a glimmer of something I hadn't noticed at the time. I check my watch: 3:59:38 . . .

39 . . . 40. I hear children's laughter; 3:59:45 . . . 46 . . . 47. I feel a curious tension as the moment of decision approaches . . . 52 . . . 53 . . . 54 . . . I still don't know what I'm going to do . . . 58 . . . 59 . . . Decision: stay put. I listen to the trees. I look at my watch: 4:00:27. *Oh, sod this!* I think, and I'm up and on my way. *I have heard your words.*

Were they acts of free will? Not getting up from the bench at 4:00? Breaking my self-imposed rule to stay put till 4:01? Did human consciousness thus intervene in the blind, deterministic, cause-and-effect machinery of the physical universe? Did I not determine something to happen? The physical act of rising from a seated position (equal and opposite forces, etc.) obeys Newton's laws to the letter, to the millionth decimal place. But what was the trigger? My free will caused a change in the physical world at a particular time and place without need of prior causes, didn't it? No it didn't. It simply can't. It just feels that way.

A FAT DRUNK plonks himself down next to me. Why me? Why my bench? There are plenty of free benches. He pulls a can of Special Brew from a supermarket bag. He's had a few already, I'd say. "This is the one," he says. "This universe. This is the one." He is smiling like the Buddha. He offers me a can of lager. I decline. "It's OK," he says, "I won't detain you. As long as I know this is actually the one. Phew! That's all right, then." I leave him smiling his Buddha smile with the sun sinking over the estuary.

I DROP BY at the co-op. It would be exciting to steal something but I haven't the nerve. I buy fish, rice and wine and head home, stopping to stare through the large window of the funeral office next door, a place now as deeply ingrained in my memory as any other. I feel no sharp emotions, just dull ones. I see myself sitting on the too-grand sofa waiting to collect Kate's ashes. They came in a green plastic container like a small milk churn. Bigger than I expected. I put it in my rucksack and set off up the hill. It was an October day much like this. I felt the intimate cargo pressing at my back but the rest of me was weightless. I floated up the hill like an autumn leaf on the breeze.

Apophany 2

KATE'S WEDDING RING HAS TURNED UP. IT DOESN'T SEEM quite right to say I found it. More a case of it presenting itself. I've showered and gone to the bedroom to get some clothes. I open my T-shirt drawer and there, lying on top of the stack of folded T-shirts, is the ring. Right on top and right in the middle, like it's a museum exhibit. I can't believe my eyes. I use that drawer every day. I've worn and washed and replaced the stuff in the drawer on a regular basis over the past nine months since the ring went missing. Every day pretty much. So it's bizarre that it should just turn up like that. But it gets weirder. Just as I'm finding the ring Fay calls from the next room. A box of photographs has fallen from the top of a wardrobe and spilled a single picture. A picture of Kate.

Now that was one of those apophanous moments. Where you simply can't help but read something magical, something supernatural, into a coincidence. It felt odd enough that the ring would just turn up like that, but to have a photo of Kate flying down from the top of the wardrobe at the moment of discovery beggars belief.

OK, I said to Fay, you've set this up, haven't you? You found the ring somewhere, I don't know, in a crack in the floorboards or somewhere, and you set it up so I would be the one who found it, and the photo, well, you just threw that in for good measure. No, she said, she certainly had not done that. She reminded me of the time when she really thought she might have found the ring, and she'd got excited about it and come straightaway to show me, but it wasn't the one. It was a different lost ring. It was the Latvian *Namejs* ring I'd bought on my first trip to Riga with my father back in the 1990s. (She'd found it the week my dad died, which was nice enough timing.) No, no, no, absolutely, no, she had not set things up. Yes you did, I said. *NO I* DID NOT! she bellowed back.

So I'll just have to accept that it was a mysterious coincidence, won't I? The ring had been in the drawer all the while, maybe lodged in a crevice, and with the traffic of clothes it somehow worked its way free over the months and plonked itself on top of the pile, right in the middle of a freshly washed and folded T-shirt. The photo box toppled off the wardrobe, as things are apt to do from time to time, and it just happened to be *at exactly the same time* as the ring reappeared, and it could just as well have been a picture of somebody else that spilled out, couldn't it?

It could.

Into the Labyrinth

WHAT A DISPIRITING PLACE IS GLASTONBURY, WITH ITS Gothickry, its plastic witchcraftery and New Age quackery. The High Street shops with their pick 'n' mix magic and mythical tack. The sickly smells of incense and godknowswhat. Slit my throat if you catch me in a purple robe.

That was the gist of the conversation. We finished our coffee and walked the half-mile or so out of town to Glastonbury Tor, the conical hill that rises from the Somerset Levels like a child's crayon drawing. Fay thought it didn't look real, by which she meant it looked man-made because of the sculpted ridges, now clearly visible in the early evening sunshine as we skirted the northern slope. There are seven deep terraces encircling the hill and no one knows for sure who cut them and for what purpose. According to one theory, they are the remnants of a prehistoric labyrinth.

At the foot of the Tor Fay's phone rang and her apologetic wince told me straightaway who it was. She raised a finger and mouthed *one minute*. I shrugged my disapproval and went to sit in the shade of a tree as Fay, out of earshot, paced the field in conversation with her ex. He'd called yesterday. He was becoming a nuisance. Five minutes passed, ten minutes, so I got up and started my ascent of the Tor alone. Stopping to catch my breath halfway up, I could see Fay, now minuscule, still pacing and talking. He's crying, he's hopeless, she said, when finally she joined me. You reap what you sow, was my reply. I had never met the man. All I knew, and I wasn't inclined to delve too deep, was Fay's account of a charming, intelligent, good-looking lawyer called

Nick, who turned out also to be manipulative, abusive and hand-in-glove with the villains who crossed his professional path. "He's a psychopath." "Do psychopaths cry?" "If it gets them what they want." "He wants forgiveness." "Forgive him, then, and tell him to fuck off." We sat in silence for a long while looking out across the Summerland Meadows. Fay took my arm and rested her head against my shoulder.

"If only I'd met you before I met him. Imagine."

"I was happily married to someone else."

"What if we'd met before you met Kate?"

"You weren't even born."

The terraces were more evident to our left on the shaded north side of the hill. As far as I could tell we were sitting in the fourth circle of the labyrinth. The Glastonbury labyrinth theory is relatively new. In the 1940s an Irish writer called Geoffrey Russell suppos-edly had a profound mystical experience, feeling himself absorbed into a pattern of concentric circles, which he somehow understood to be identified with the workings of his brain. He made sketches of his visions and, much later, in the 1960s, came across photo-graphs of two Cornish rock carvings of circular labyrinths that bore a striking resemblance to his drawings. He believed he saw the same pattern in aerial photographs of Glastonbury Tor. The designs of his brain, it seemed, were etched in Cornish slate and marked the Somerset landscape.

I've been doing my research. The earliest-known representa-tion of a labyrinth is carved on a piece of mammoth ivory discov-ered in a Paleolithic tomb in Siberia. It's seven thousand years old, at least. You find similar images across the world from Mexico to India, North America to Spain. The seven-coil labyrinth that ap-pears on the coins of ancient Greece links directly to the Bronze Age civilization of Crete and, an ocean and an age away, the same design is sacred to the Native American Hopi people. Their name for it is Mother and Child. It symbolizes the tribe's emergence from Mother Earth at the beginning of time and the everyday wonder of childbirth. In India they gave a similar image to pregnant women, who would trace the path outward from the center in order to focus the mind on the task of labor and childbirth. The seven-circuit symbol is found carved on rock, cut into turf, woven in fabric, laid out in stones on the shoreline, and set in tiles on the floors of churches and cathedrals. It's a path for fingers or feet, or for the eye to follow, taking with it the mind on a looping trail to the interior.

"So we were both having a miserable time," Fay said. "You with your dying wife, and me with my nasty Nick."

"Miserable?"

I thought about that. *Miserable?* I suppose it was pretty wretched at times but, no, those eight years from diagnosis to death were not a time of unremitting misery. There was plenty of common happiness, some moments of real joy, and life between the scans and the treatments and the hospital consultations settled mostly into ordinariness. It got grim in the final year, with the cancer drilling deep into Kate's bones and lungs, and lymphedema disfiguring her body. Cancer is a cruel and merciless disease. But you do what you can. There's consolation in doing what you can. *Miserable?* The last days were overwhelmingly sad, but that's not the same, and, even then, there were moments that now sparkle as jewels of memory. Here's one. Five days before Kate died we went up to Plymouth Hoe. It was her last excursion out of the house. She was in a wheelchair and needed an oxygen cylinder. We sat looking out on the dazzling waters of the Sound, not saying very much, and she managed to dispense with her oxygen tubes for a while to relish the sea air. "This is beautiful," she said, and then, brightly, "I won't be here next week." Bizarrely, we were both smiling.

"Miserable?"

"But your misery was more wholesome than mine."

What was I smiling at now? I was smiling at *wholesome.* When Kate was entering the end stage of her illness several of our friends were having relationship troubles, some to the point of separation and divorce. I found myself drawn into candid conversation, usually with the men over a drink, but with one or two of the women as well. They were all sympathetic to my predicament, of course, as I was to theirs. What you two are going through is so much worse, they would say, but I couldn't help feeling that, for some of them, there was an element of envy. My relationship with Kate was about to reach its inevitable conclusion, but it would be a clean, final break, unlike the guilt-ridden tearing apart they were going through. They thought that my liberation—not that I was looking for "liberation"—would be more honorable, more complete than theirs. I thought of this as *bereavement envy.*

Kate never read any philosophy, or anything other than fiction for that matter, not for pleasure anyway. She was a voracious reader of novels, one or two a week compared to my one or two a year, but had no time for other stuff. I recommended the Stoics once and she gave me a weary smile. Truth was, she was bearing up to her suffering as stoically as Marcus Aurelius, Seneca, or any of the others ever could have. Death held no fear. None whatsoever. It was life that mattered, what was left of it. The last dregs were to be savored. *You don't know how precious life is. You think you do, but you don't.*

The day after our excursion to the Hoe, a Sunday, we celebrated our granddaughter Millie's second birthday. It was a milestone that had to be reached. Kate wasn't well enough to come downstairs so we all went up to the bedroom and sang "Happy Birthday Dear Millie" along to the jingle of a musical birthday card. Two days later it was Kate's birthday, her fifty-seventh. What gift do you get for someone on the very brink of death? Nat and Rosie got her a cashmere blanket. I wrapped it around her when I settled her for the night. She whispered, "This would be a good day to go."

THE UNIVERSAL FASCINATION with the image of the labyrinth suggests some fundamental psychological significance, that perhaps it holds the power to captivate and transform the mind in some way. It's been suggested, for example, that threading the spirals of a labyrinth works to loosen the grip of rational, analytical, "left-brain" styles of thinking, thereby opening up the mind to more intuitive, spiritual, "right-brain" modes of experience, to the language of poetry and the imaginal reality of ghosts and gods. The labyrinth may be linked to consciousness itself. The psychologist Nicholas Humphrey speculates that the recurrence of spiral imagery in prehistoric art signifies the evolution of self-awareness, or, as he puts it, *the arrival of souls on the scene.* Human beings were not only sentient (like other animals) but aware that they were aware. They had come to understand the interiority, the inviolable *privacy*, of mental life. Humphrey was taken aback by the remarkable resemblance between a figure he had drawn for a scientific journal and a 17,000-year-old cave painting he later saw on a visit to Vilafamés in Spain. Both are schematic representations of the human form with near-identical spirals coiling into the head. The scientific drawing was one of a series illustrating the gradual "privatization of experience" through the course of evolution, and it occurred to Humphrey that possibly, just possibly, the cave artist was also striving to capture the interiority of mental life.

Carl Jung, the founder of analytical psychology, would have loved these observations. He thought the labyrinth was a primordial image of the psyche. It symbolized the process of *individuation*, the winding, snakelike path to psychological wholeness and authenticity. Beset with traps and terrors, the path descends deep beneath the surface of conscious selfhood to the secret chambers of the unconscious mind. If we are to achieve wholeness we must summon courage to endure the darkness and dread of the psychic labyrinth. We must confront the monsters that lurk within, and then, having reached our goal, the center of the self, the true self, we must find our way back to the outer world. The dangers are real. To become lost in the maze is to succumb to the demons of depression or anxiety, or even psychosis.

WE GOT TO the top of the Tor and shared a bottle of water in the cool of St. Michael's Tower, the roofless remnant of a fourteenth-century church. Then we sat out on the grass, or rather Fay did. I'd left my rucksack in the tower and went to retrieve it. "Looking for this?" My heart kicked. He seemed to spring from nowhere. But this is Glastonbury. Do not be too surprised by the sudden materialization of a fat man with large white wings and a halo. He had my rucksack in one hand and a can of Special Brew in the other. I took the rucksack. The fat man proffered a damp hand. "St. Michael," he said, "the Archangel." His wings were impressively luxuriant and I couldn't figure how he'd pulled the halo trick, presumably some arrangement of lights in his collar, though I couldn't see any. "Mike. Remember?" Yes, indeed I did. The man who came tapping at my window. Paddington Station. Yes. I remember. Trebarwith Bay. Devonport Park. Mike the drunk. Mike the time-twister. Things were falling into place.

"Welcome to the heart of the labyrinth."

"Well," I said, "fancy seeing you here."

You know that I know that you know that this is no coincidence, he signaled, wordlessly. "Have one of these." He retrieved another can of strong lager from a plastic supermarket bag. I declined, but he opened it anyway and stuck it in my hand. I told him if this was the heart of the labyrinth, then he wasn't much of a monster. Not exactly the Minotaur.

MISERABLE? They were hard, hard, hard those last months. They were heartbreaking. You come to embrace hopelessness. There is no hope. Hope is worthless. Hope is delusional. Hope is counterproductive. Accept the hopeless truth. You have to get beyond hope. See hopelessness as an honest companion who walks the last days with you. Hopelessness gets you through the day: the arsewiping, the medication, the lymphedema massage. We did the dismal massage thing every day, endlessly creaming and kneading, and it made no difference. We did it forlornly, joylessly, hopelessly, but love was also part of the process. Love outlives hope. Get hope out of the way.

So, MIKE THE Archangel, what do you want?"

"What if you could go back and change things? You go back in time and you're given a choice. It's up to you whether or not your wife gets cancer."

"Hmm. Tough call!"

"No. Seriously. What if?"

I could see Fay out on the grass. Enough of this fat drunk and his fantasies. I turned to go. He grabbed my arm. "You would, wouldn't you? But there's a condition. Remember what C. S. Lewis told you? If he could have cured his wife's cancer by arranging never to see her again, then that's what he would have done. Well, would you do that?"

"Why don't you just piss off and leave me alone?" I said.

I was on my way but something compelled me to answer his question. I told him that since my encounter with Lewis I had, in fact, already given the matter some thought and it would be a very tough choice. What would the arrangement be, exactly? Would Kate be in the picture? Would it be something we'd both agreed to, or would I just have to slope off somewhere in secret without her ever knowing the reason why? What about the children? Would I see them again? There were so many complications.

"Does that answer your question?"

A young Japanese couple had entered the tower and stood for a while staring up through the open roof at the fading sky and wispy clouds.

"Anyway," I said, "I'm glad it's a choice I don't really have to make." The fat man drew close. With lowered voice and beery breath, he said, "But you do."

So he made his offer. It wasn't the C. S. Lewis scenario. It was something more radical, simpler, cleaner and without the complications of a mysterious disappearance or the pain of a negotiated separation. I told him it was an interesting proposition. It raised a lot of questions. I would have to think it through, as I'd contemplated Lewis's thought experiment. Yes, I said, edging away, let me think it through. I wasn't taking him seriously, he said.

Next thing I know we're standing at the entryway of the tower looking eastward over the hazy green fields of Somerset, a buzzard wheeling high above.

Look!

I looked and the buzzard stopped wheeling. Just stopped, its broad wings spread motionless, frozen in the sky. The people around us were all as still as snapshots. A dog was locked, mid-air, as it leapt for a ball. The fixed, contorted face of a crying child intensified the silence. A strand of Fay's hair had been lifted by a breeze and just hung there. There was no breeze now, just odd variations in the texture of the still air as I waved my hand through it.

"So, you send me back to a time before Kate and I met, to a choice-point. I am—what?—some sort of amalgam of my old—that's to say, young—self and my present self. I innocently, yet somehow knowingly, take a decision that changes the course of my life, and Kate's. It's a life in which she doesn't get cancer and die a young death, but it's a life in which I play no part. We never get to know each other."

Mike the Archangel smiles. His wings beat and his smiling face floats up through the tower and into the sky. I watch it recede to a pinpoint and disappear into the stratosphere. I look down and I'm in a pub. I recognize the place, though can't remember the name. I'm finishing my pint and I'm going to catch a bus into town to see a girl I met at the college disco last night. True to her name, Melanie, she has jet-black hair.

"Same again?" This is my friend Rob.

"No, I've got a date."

"Half?"

Well, I could, I suppose. I've got ten minutes. I could knock one back and dash for the bus. I picture Melanie: glossy black hair, tall and slender, pretty face, and I see myself on the bus heading down through the suburbs and into the city. Then I'm outside the Crazy Daisy nightclub. Melanie, according to convention, will be a few minutes late. Then we might go into the club, or we might go somewhere else. I don't really like nightclubs. It was her idea.

The angel-drunk is speaking now, from somewhere:

"Try not to be sentimental. It was pure chance you met Kate. It's pure chance anybody meets anybody. When you turned up late at the club and your girl wasn't there, you might have just gone off to try your luck somewhere else rather than go inside and chance upon meeting Kate."

"What do you mean, 'sentimental'?"

"You know, dwelling on the good times you and Kate had together, the children and all that. Don't worry about that. You'd have made another life. You *could* make another life, with someone else, starting from scratch. There are so many roads untraveled, other lives and different experiences, and other precious children to love, if that's what you want. There's an *infinity* of other lives. Now's your chance."

I hear laughter and the snap of a ring-pull, "It's all so fucking *random*!"

Rob speaks: "Same again?"

"No, I've got a date."

"Half?"

My thoughts swirl in the smoke of the saloon bar. I feel the future. It's rich. It's hard. It's good. It's like a million futures. If you

want a life, that's what you get as often as not. The angel is somewhere in the cigarette smoke.

"So, I have that drink and I miss the bus and my life follows
the same course."

"Every breath, every blink."

"But if I don't have the drink . . ."

"You catch the bus to a different life."

"I never get together with Kate, but she stays well and lives a
long life?"

"That's the deal."

"And all memory of the life I've had is wiped from my brain,
including you."

The smoke thinned.

Rob: "Same again?"

"No, I've got a date."

"Half?"

The fat man was right about the randomness of it all. We have
no say in being born, and the fact that we exist at all is a matter of
pure and improbable accident: those particular parents, that particular egg, that particular sperm, a glass of beer, a missed bus,
and a trillion throws of the dice in between. There are many possible lives to live in the labyrinth of time but how could I throw
away the one I've had? A memory intrudes. I see myself striding
into cold surf, clutching a precious little object, a silver dolphin.
I swim far out, far enough to feel a current of fear when I think
about the depth of the water. The human figures on the beach are
microscopic. Kate and the boys are there somewhere but I can't
make them out from the other microbes. I turn to the horizon and
hurl the silver dolphin as far as I can.

"You wouldn't be throwing anything away. Listen. There's a
short story by Arthur C. Clarke, a very short story, just thirty-one
words, and it's about God *unmaking* the Universe; 'siseneG' it's
called—that's 'Genesis' backwards. He deletes lines of the computer code that runs the Universe and the Universe ceases to exist.

Then he ponders for a while, a few aeons, and adds: 'ERASE.' And now the Universe never *had* existed. See? You can, if you want, totally erase the life you've had. It will never have existed."

"But it happened!"

"It won't have."

The thought horrifies me. "But how can something that's already happened be made to *un*happen?"

"Because the past is indeterminate."

I remember something he'd said that time we met on the train. "You told me all eternity, past, present and future, was fixed in a space-time block. Everything that has happened and everything that will happen is already locked in place. The passage of time is an illusion."

"It is."

"So, if everything that ever happens is already set in place, how can you go back in time and change the course of events?"

"You can't, not if you restrict your options to a single universe. But our cosmos, our space-time manifold, is one of an infinite number of universes. Everything that can possibly happen will happen, over and over again, ad infinitum. Every possible history will play out in an infinite number of universes. There are innumerable places with histories identical to our own, right down to the atomic level, and places with variations on the theme, major and minor. So there are infinite histories to choose from with infinite variations on the theme of you and your life. It's just a matter of . . ."

"No, no, no, no, don't bother! God, NO!"

Same again?

No, I've got a date.

Half?

Am I being selfish, I wonder, clinging to the life I've had at the cost of Kate's suffering? That's a fate I could free her from.

"*All* memories erased?"

"You can keep one," says the fat man. "It will come to you as a recurring dream, once a year. You won't quite be able to make sense of it, not explicit sense anyway, but it will mean something to you all the same. It'll make mythical sense."

As he spoke, a memory came to mind. The day before the ultimate Bad News Consultation with the oncologist, Kate and I drove over to Bigbury-on-Sea. We didn't know it was going to be such bad news, but bad news was coming sooner or later. By this time she was in a lot of pain and discomfort with the bone tumors, her lungs were failing and her legs and hips were swollen with lymphedema. But she was determined to have one last swim in the sea. The truth is I was not looking forward to it. We had the wheelchair but that was no use with the narrow steps from the car park to the beach to negotiate. I wasn't sure she'd make it with just a crutch and me to bear her weight. I did not want her to be embarrassed. I did not want me to be embarrassed. But she did make it. It took a while but we got to the bottom of the steps and I erected a canvas chair for her to sit on. And there she sat, facing the sun, while I read a newspaper. There was a family not far away, well spoken and with an air of superiority about them. They were down from London. The pretty, skinny, teenaged girl looked miserable. Whatever else might be troubling her I sensed she was displeased to have us settle so close by when there was plenty of space along the beach, but we weren't going any further. Then, when the sea crept closer, we had our swim. Kate was wearing a swimsuit but was not going to expose her lymphedema-battered body, especially not to the sullen teenager, so she hobbled into the sea still wearing her rose-print summer dress, and she glided out through the cold water, graceful again, and joyful. We both knew this was the last time, though neither of us voiced the thought. It felt to me as if it was as good a moment as we'd ever had.

Same again?

No, I've got a date.

Half?

I made my choice. It came down to a *feeling of rightness*.

I'D LOST MYSELF for a while, staring up through the gaping roof, alone in the tower. Fay was calling. *It's beautiful*, she said, *look*, and there was the rising Moon, as good as full. She said it was a blue Moon, or would be tomorrow.

Further Reading

LISTED IN THE Select Bibliography are some of the works I consulted during the writing of this book. They are presented without commentary, but first I want to pick out some volumes for those of you who might feel energized to dig specifically into matters of mind, brain and consciousness. Christopher Frith's *Making Up the Mind: How the Brain Creates our Mental World* is a lucid and entertaining introduction to cognitive neuroscience (with an emphasis on perception and action). If you are new to questions of brain and mind, it's a good place to start. Professor Frith was also a consultant for Rita Carter's *The Brain Book*, a beautifully illustrated beginners' guide. There are many excellent books geared for undergraduate studies in neuropsychology (and behavioral neuroscience more generally), but if I were to pick just a couple they would be G. Neil Martin's *Human Neuropsychology* and Kolb and Whishaw's *Fundamentals of Human Neuropsychology*.

I want also to recommend some rather more heavy-duty texts for those who might truly wish to immerse themselves in the waters of clinical neuropsychology. The evening before the interview for my first job in the field I sat in a damp room at a B&B in a seedy part of Leeds, opened a bottle of beer and leafed through the second edition of Muriel Lezak's *Neuropsychological Assessment*. It illustrated brilliantly the essential convergence of theory and practice that lies at the heart of clinical neuropsychology and convinced me that this really was a line of work that I could launch myself into and do well at. Now into its eighth edition, and multi-authored, it remains a key text. I can also recommend the compendious *Handbook of Clinical Neuropsychology* (Gurd et al., eds.).

The best basic introduction to consciousness studies is Susan Blackmore's *Consciousness: An Introduction*. Her collection of interviews with leading figures in the field, *Conversations on Consciousness*, is also recommended. It's nigh on three decades since Daniel Dennett wrote the much misunderstood, and often misrepresented, *Consciousness Explained*. It stands the test of time. The best introduction to consciousness from a neurological perspective is Adam Zeman's *Consciousness: A User's Guide*. Nicholas Humphrey has been a significant influence on my thinking about consciousness in recent years. All his books are worth reading and I could recommend a number of his works, but the book resting on the windowsill in front of me right now is his most recent, *Soul Dust*, which is as good a place to start as any. Michael Gazzaniga has been an inspiration down the years and his recent scientific autobiography, *Tales from Both Sides of the Brain: A Life in Neuroscience*, gives a brilliant insight into the life and work of a pioneer brain scientist.

Select Bibliography

Aldrin, B. and Abraham, K., *Magnificent Desolation: The Long Journey Home from the Moon*. New York: Crown, 2009.

Blackmore, S., *Consciousness: An Introduction*. London: Hodder Education, 2010.

Blackmore, S., *Conversations on Consciousness*. Oxford, UK: Oxford University Press, 2010.

Camus, A., *The Myth of Sisyphus*, trans. Justin O'Brien. London: Penguin, 2000.

Carter, R., *The Brain Book*, 2nd ed. London: Dorling Kindersley, 2014.

Cave, S., *Immortality: The Quest to Live Forever and How it Drives Civilisation*. London: Biteback Publishing, 2012.

Clark, A., *Being There: Putting Brain, Body and World Together Again*. Cambridge, Mass.: MIT Press, 1998.

Clarke, A. C., "siseneG," *Analog* magazine, May 1984.

Damasio, A., *Self Comes to Mind: Constructing the Conscious Brain*. London: William Heinemann, 2010.

Dennett, D., *Consciousness Explained*. London: Penguin, 1993.

Descartes, René, *Meditations on First Philosophy*, trans. D. M. Clarke. London: Penguin, 2000 (orig. publ. 1641).

Feinberg, T. E., *From Axons to Identity: Neurological Explorations of the Nature of the Self*. New York: W. W. Norton and Company, 2009.

French, C. C. and Stone, A., *Anomalistic Psychology*. London: Palgrave Macmillan, 2014.

Frith, C. D., *Making Up the Mind: How the Brain Creates our Mental World*. Oxford, UK: Blackwell Publishing, 2007.

Gazzaniga, M. S., *Tales from Both Sides of the Brain: A Life in Neuroscience*. New York: HarperCollins, 2015.

Gleick, J., *Time Travel*. London: HarperCollins, 2016.

Grayling, A. C., *Descartes: The Life of René Descartes and Its Place in His Times*. London: Pocket Books, 2006.

Gurd, J., Kischka, U. and Marshall, J., *The Handbook of Clinical Neuropsychology*, 2nd ed. Oxford, UK: Oxford University Press, 2010.

Harding, D. E., *On Having No Head: Zen and the Discovery of the Obvious*. Carlsbad, Calif.: Inner Directions, 2002.

Hawking, S. and Mlodinow, L., *The Grand Design: New Answers to the Ultimate Questions of Life*. New York: Random House, Inc., 2010.

Herodotus, *The Histories*, intro. J. M. Marincola, trans. Aubrey de Sélincourt. London: Penguin, 2003.

Hesiod, *Theogony* and *Works and Days*, trans. M. L. West. Oxford, UK: Oxford University Press, 2008.

Hillman, J., *Pan and the Nightmare*, 2nd ed. Woodstock, Conn.: Spring Publications, 2007.

Hofstadter, D., *I Am a Strange Loop*. New York: Basic Books, 2007.

Homer, *The Iliad*, intro. B. Graziosi, trans. A. Verity. Oxford, UK: Oxford University Press, 2012.

Humphrey, N., *Seeing Red: A Study in Consciousness*. Cambridge, Mass.: Harvard University Press, 2006.

Humphrey, N., *Soul Dust: The Magic of Consciousness*. Princeton, N.J.: Princeton University Press, 2011.

Hurley, S. L., *Consciousness in Action*. Cambridge, Mass.: Harvard University Press, 1998.

Irvine, W., *A Guide to the Good Life: The Ancient Art of Stoic Joy*. Oxford, UK: Oxford University Press, 2009.

Jaynes, J., *The Origins of Consciousness in the Breakdown of the Bicameral Mind*. Boston, Mass.: Houghton Mifflin, 1977.

Kennaway, J., "Singing the Body Electric: Nervous Music and Sexuality in Fin de Siècle Literature," in A. Stiles (ed.), *Neurology and Literature, 1860–1920*. London: Palgrave Macmillan, 2007.

Kershaw, S. P., *A Brief Guide to the Greek Myths*. London: Robinson, 2007.

Kolb, B. and Whishaw, I., *Fundamentals of Human Neuropsychology*, 7th ed. New York: Worth Publishers, 2015.

Le Doux, J., *Synaptic Self: How Our Brains Become Who We Are*. London: Penguin, 2003.

Lewis, C. S., *A Grief Observed*. London: Faber & Faber, 1961.

Lezak, M. D., Howieson, D. B., Bigler, E. D. and Tranel, D., *Neuropsy-chological Assessment*, 5th ed. Oxford, UK: Oxford University Press, 2012.

MacDonald, P. S., *History of the Concept of Mind: Speculations about Soul, Mind and Spirit from Homer to Hume*, vol. 1. Farnham: Ashgate Publishing, 2003.

McGinn, C. *The Mysterious Flame: Conscious Minds in a Material World.* Oxford, UK: Blackwell Publishing, 1999.

McGrath, A., *C. S. Lewis—A Life: Eccentric Genius, Reluctant Prophet.* London: Hodder & Stoughton, 2013.

Makari, G., *Soul Machine: The Invention of the Modern Mind.* New York: Norton & Co. Inc., 2015.

March, J., *The Penguin Book of Classical Myths.* London: Penguin Books, 2008.

Marcus Aurelius, *Meditations*, trans. Martin Hammond. London: Penguin Classics, 2006.

Martin, G. N., *Human Neuropsychology*, 2nd ed. Harlow, UK: Pearson Education Ltd., 2006.

Martino, P. and Milkowski, B., *Here and Now! The Autobiography of Pat Martino.* Milwaukee, Wisc.: Backbeat Books, 2011.

Matyszak, P., *The Greek and Roman Myths: A Guide to the Classical Stories.* London: Thames & Hudson, 2010.

Mitchell, E., *The Way of the Explorer: An Apollo Astronaut's Journey Through the Material and Mystical Worlds.* New York: Putnam, 1996.

Nietzsche, F., *A Nietzsche Reader*, intro. & trans. R. J. Hollingdale. London: Penguin, 2003.

Noë, A., *Out of Our Heads: Why You Are Not Your Brain, and Other Lessons from the Biology of Consciousness.* New York: Hill and Wang, 2009.

Noë, A., *Strange Tools: Art and Human Nature.* New York: Hill and Wang, 2015.

O'Regan, J. K., *Why Red Doesn't Sound Like a Bell: Understanding the Feel of Consciousness.* Oxford, UK: Oxford University Press, 2011.

Parfit, D., *Reasons and Persons.* Oxford, UK: Oxford University Press, 1984.

Plutarch, "The Life of Theseus," trans. Bernadotte Perrin, in *Delphi Complete Works of Plutarch.* Hastings: Delphi Publishing, 2013 (Kindle edition).

Ramachandran, V. S., *The Tell-Tale Brain: Unlocking the Mystery of Human Nature*. New York: Windmill Books, 2011.

Ridpath, I., *Star Tales*. Cambridge, UK: Lutterworth Press, 1988.

Sayer, G., *Jack: A life of C. S. Lewis*. London: Hodder & Stoughton, 1997.

Schopenhauer, A., *Essays and Aphorisms*, trans. R. J. Hollingdale. London: Penguin, 1976.

Seneca, *Letters from a Stoic*, trans. Robin Campbell. London: Penguin, 2004.

Seneca, *On the Shortness of Life*, trans. C. D. N. Costa. London: Penguin, 2004.

Sharpless, B. and Dhogramji, K., *Sleep Paralysis: Historical, Psychological and Medical Perspectives*. Oxford, UK: Oxford University Press, 2015.

Sibley, B., *Shadowlands: The True Story of C. S. Lewis and Joy Davidman*. London: Hodder & Stoughton, 2013.

Smith, R., *Between Mind and Nature: A History of Psychology*. London: Reaktion Books, 2013.

Snell, B., *The Discovery of the Mind in Greek Philosophy and Literature*. New York: Dover Publications, 1982.

Sperry, R. W., "Hemisphere Disconnection and Unity in Conscious Awareness," *American Psychologist*, vol. 23 (10), 1968.

Stroud, R., *The Book of the Moon*. London: Doubleday, 2009.

Velmans, M., *Understanding Consciousness*. London: Routledge, 2000.

Waterfield, R. and Waterfield, K., *The Greek Myths: Stories of the Greek Gods and Heroes Vividly Retold*. London: Quercus Publishing, 2011.

Zeman, A., *Consciousness: A User's Guide*. New Haven, Conn.: Yale University Press, 2003.

Acknowledgments

I WAS PRIVILEGED to spend a good part of my career working for the UK's National Health Service and it is a privilege now to be writing about some of the remarkable people I encountered along the way. They wouldn't know, but I learned so much from them. Names have been changed, of course, and layers of disguise have been added to preserve anonymity, but my case stories are very much based on real people. If it were possible to thank them in person, I would.

The autobiographical elements of the book (once disentangled from the obvious fictional threads) are also true, although, here too, some names have been changed. By coating real events with this thinnest veneer of fiction, I hoped to create some small emotional distance from events and thus, perhaps, obtain a clearer view. But this, I confess, is a post hoc explanation and I don't really know how true it is. In fact, I think I just made that up. Anyhow, my wife's name was Sonja, not "Kate," as she appears in these pages. That's the name she chose for herself in my previous book, *Into the Silent Land*, so I stuck with it. And thanks to Dan, Charlotte, Jon, Pete and Fiona for allowing versions of themselves to appear in this odd quasi-fictional form. Also, Ebony, Millie and Harry, who made very clear they wanted to appear as themselves.

This book was not an easy ride for my editors, so sincere thanks to them all. At Penguin: Helen Conford and Chloe Campbell. Helen, who commissioned the book, has been patient beyond the call of duty. Charlotte Ridings read the full text with an eagle eye and corrected numerous infelicities. At Crown: Kevin Doughten, Jon Darga and Alex Hamilton. Kevin, especially, has been a champion. As well as these talented professionals, I also have a talented amateur to thank for a significant editorial contribution. Fiona Walker read different versions of the text with a judicious and creative eye. This would, quite simply, have been a different book without Fi, not least because she found her way into it.

There are numerous people who have, in different ways, helped shape the content of the book. I've had many enjoyable and instructive conversations with Adam Zeman on and around the topic of consciousness and I thank him, in particular, for inviting me to interview his patient, "Graham," who features in "Rotten to the Core." Another friend, the filmmaker Ian Knox, introduced me to Pat Martino ("Tabula Rasa"). Some years ago I had the pleasure of working with Ian on a documentary film about Pat's extraordinary life, *Martino Unstrung* (Dir. Ian Knox, Sixteen Films, 2008), and I have drawn heavily on experiences absorbed during the making of the film. Pat's autobiography, *Here and Now!*, co-written with Bill Milkowski, was a valuable source of information on the details of Pat's early career, and conversations with Victor Schermer at All About Jazz helped sharpen my thinking about the neuropsychology as well as the music of Pat Martino. And, of course, I am hugely grateful to the remarkable Mr. Martino himself for his time, generosity and energy, both during the making of the film and in our times of friendship since.

Thanks also to another filmmaker, Carla MacKinnon, who was generous in allowing access to her files on sleep paralysis and in sharing her own experiences of that disturbing condition. I am grateful to Martin Rees, Astronomer Royal, who took time out of a very busy schedule for a lengthy interview, parts of which were used in a piece I wrote for *Prospect* magazine. I have extracted a few snippets of our conversation and made use of them in "Aliens." Heartfelt thanks to Bharat Maldé for sharing his insights into the experience of bereavement and for permission to write about the death of his beloved son, Anjool, in "Living the Dream." Suzanne Wilson is the psychotherapist friend whose words I quote in that chapter. Claire Delle Luche advised with the French in "*Épater les Bourgeois*," but I blame her for slowing the progress of the book in its early stages. Thanks for the music, Claire. Happy days.

This would be a lesser book without Garry Kennard's atmospheric drawings. They are just a subset of a larger suite of images relating to the themes of the book, and the plan, in due course, is to exhibit them in full. For further examples of Garry's work, visit garrykennard.com.

Several pieces within this collection borrow from material previously aired in *Prospect*, *The Times*, the *Guardian*, *New Scientist* and *The Psychologist*, and I am grateful to my editors at those publications for opportunities to write and develop ideas.

Warmest thanks to my agent, Rebecca Carter at Janklow & Nesbit, and her predecessor Claire Paterson, for their fabulous encouragement and support from the beginning to the end of the project, and for providing many valuable insights along the way.